Catalysis and Surface Properties
of Liquid Metals and Alloys

CHEMICAL INDUSTRIES

A Series of Reference Books and Textbooks

Consulting Editor
HEINZ HEINEMANN
*Heinz Heinemann, Inc.,
Berkeley, California*

Volume 1: Fluid Catalytic Cracking with Zeolite Catalysts, *Paul B. Venuto and E. Thomas Habib, Jr.*

Volume 2: Ethylene: Keystone to the Petrochemical Industry, *Ludwig Kniel, Olaf Winter, and Karl Stork*

Volume 3: The Chemistry and Technology of Petroleum, *James G. Speight*

Volume 4: The Desulfurization of Heavy Oils and Residua, *James G. Speight*

Volume 5: Catalysis of Organic Reactions, *edited by William R. Moser*

Volume 6: Acetylene-Based Chemicals from Coal and Other Natural Resources, *Robert J. Tedeschi*

Volume 7: Chemically Resistant Masonry, *Walter Lee Sheppard, Jr.*

Volume 8: Compressors and Expanders: Selection and Application for the Process Industry, *Heinz P. Bloch, Joseph A. Cameron, Frank M. Danowski, Jr., Ralph James, Jr., Judson S. Swearingen, and Marilyn E. Weightman*

Volume 9: Metering Pumps: Selection and Application, *James P. Poynton*

Volume 10: Hydrocarbons from Methanol, *Clarence D. Chang*

Volume 11: Foam Flotation: Theory and Applications, *Ann N. Clarke and David J. Wilson*

Volume 12: The Chemistry and Technology of Coal, *James G. Speight*

Volume 13: Pneumatic and Hydraulic Conveying of Solids, *O. A. Williams*

Volume 14: Catalyst Manufacture: Laboratory and Commercial Preparations, *Alvin B. Stiles*

Volume 15: Characterization of Heterogeneous Catalysts, *edited by Francis Delannay*

Volume 16: BASIC Programs for Chemical Engineering Design, *James H. Weber*

Volume 17: Catalyst Poisoning, *L. Louis Hegedus and Robert W. McCabe*

Volume 18: Catalysis of Organic Reactions, *edited by John R. Kosak*

Volume 19: Adsorption Technology: A Step-by-Step Approach to Process Evaluation and Application, *edited by Frank L. Slejko*

Volume 20: Deactivation and Poisoning of Catalysts, *edited by Jacques Oudar and Henry Wise*

Volume 21: Catalysis and Surface Science: Developments in Chemicals from Methanol, Hydrotreating of Hydrocarbons, Catalyst Preparation, Monomers and Polymers, Photocatalysis and Photovoltaics *edited by Heinz Heinemann and Gabor A. Somorjai*

Volume 22: Catalysis of Organic Reactions, *edited by Robert L. Augustine*

Volume 23: Modern Control Techniques for the Processing Industries, *T. H. Tsai, J. W. Lane, and C. S. Lin*

Volume 24: Temperature-Programmed Reduction for Solid Materials Characterization, *Alan Jones and Brian McNicol*

Volume 25: Catalytic Cracking: Catalysts, Chemistry, and Kinetics, *Bohdan W. Wojciechowski and Avelino Corma*

Volume 26: Chemical Reaction and Reactor Engineering, *edited by J. J. Carberry and A. Varma*

Volume 27: Filtration: Principles and Practices, second edition, *edited by Michael J. Matteson and Clyde Orr*

Volume 28: Corrosion Mechanisms, *edited by Florian Mansfeld*

Volume 29: Catalysis and Surface Properties of Liquid Metals and Alloys, *Yoshisada Ogino*

Volume 30: Catalyst Deactivation, *edited by Eugene E. Petersen and Alexis T. Bell*

Additional Volumes in Preparation

Catalysis and Surface Properties of Liquid Metals and Alloys

Yoshisada Ogino

Department of Chemical Engineering
Tohoku University
Aramaki-Aoba, Sendai, Japan

Marcel Dekker, Inc.　　　　New York and Basel

Library of Congress Cataloging in Publication Data

Ogino, Yoshisada
  Catalysis and surface properties of liquid metals and alloys.

  (Chemical industries ; v. 29)
  Includes bibliographies and index.
  1. Catalysis. 2. Surface chemistry. 3. Liquid metals.
I. Title. II. Series.
QD505.O35 1987   660.2'995   86-29036
ISBN 0-8247-7699-2

Copyright © 1987 by MARCEL DEKKER, INC. All rights reserved

Neither this book nor any part may be reproduced or transmitted in any form or by any means, electronic or mechanical, including photocopying, microfilming, and recording, or by any information storage and retrieval system, without permission in writing from the publisher

MARCEL DEKKER, INC.
270 Madison Avenue, New York, New York 10016

Current Printing (last digit)
10 9 8 7 6 5 4 3 2 1

PRINTED IN THE UNITED STATES OF AMERICA

Preface

The primary intention of this book is to present up-to-date information about the catalysis and surface properties of liquid metals and liquid alloys. This book is intended for use by chemical engineers and researchers in catalysis, surface science, liquid metals, and chemical process technologies. Although several excellent monographs are available for those readers interested in the bulk properties of liquid metals and alloys, there is no other book that integrates information about their surface phenomena. Publication of this book, I believe, improves our understanding of the science and technology of liquid metals and alloys.

The catalytic properties of liquid metals and alloys are reviewed in the first four chapters. The substance of these chapters has been taken mainly from research carried out in my laboratory, because of the lack of published research by other authors. Thus, it is hoped that the information presented about experimental techniques and surface catalysis will be especially useful for those readers interested in working with liquid metals and liquid alloys. A mechanistic account of catalysis at the atomic, or electronic, level is presented in Chapter 4. The problems treated in this chapter address the structures and properties of liquid metals.

The last three chapters review the recent advances in research on the surface properties of liquid metals and alloys. The discussion covers much of the literature from 1970 through 1985. Experimental data are summarized in tables as much as possible. I would like to express my great respect to all authors who have contributed the valuable scientific information cited in this book.

Although mathematical expressions have been minimized throughout this book, this never means that rigorous theoretical treatments are unimportant. Theoretical treatments give the reader easy access to the essential knowledge accumulated in this particular field of science. To understand the present situation regarding the theory of liquid metals, readers should consult J. M. Ziman, *The Physics of Metals. 1. Electrons* (Cambridge, 1969); T. E. Faber, *An Introduction to the Theory of Liquid Metals* (Cambridge, 1972); or M. Shimoji, *Liquid Metals* (Academic Press, 1977). For readers who wish to study the theory of surface tension, the book by C. A. Croxton, *Statistical Mechanics of the Liquid Surface* (John Wiley & Sons, 1980), is recommended.

For myself, this book is a professional milestone. Thus, I express my thanks to Dr. Maurits Dekker, Marcel Dekker, Inc. who kindly recommended that I publish this book; to Carol Mayhew and Henry Boehm, editors at Marcel Dekker, Inc; for their patience and valuable advice; to Dr. Yasukatsu Tamai (Emeritus Professor, Tohoku University) for his constant encouragement; and to Dr. Sentaro Ozawa for his illustrations. I gratefully acknowledge the Ministry of Education, Science, and Culture of Japan for its constant financial support to the study on catalysis of liquid metals. I also wish to thank my family, in particular my wife Hisako, for their forbearance and hearty help during the preparation of the manuscript.

Preface

   Finally, I express my sincere personal acknowledgment to the late Dr. Hiroshi Uchida, to whom this book is dedicated.

Yoshisada Ogino

Contents

PREFACE   iii

1/   HISTORY AND SCOPE   1

    I. Introduction   1
    II. Historical Survey   2
    III. Motive for Studying the Catalysis of Liquid Metal   4
    IV. Scope of Catalyst Elements   6
       References   6

2/   TECHNIQUES FOR MEASURING CATALYTIC ACTIVITY   8

    I. Introduction   8
    II. Catalyst Purification   9
    III. Various Reactors   10
       References   23

3/   REACTIONS OVER LIQUID METALS AND ALLOYS   24

    I. Introduction   25
    II. Dehydrogenation of Alcohols   26
    III. Dehydrogenation of Amines   33
    IV. Dehydrogenation of Hydrocarbons   36
    V. Hydrogen Transfer Reactions   43

VI. Coal Liquefaction   49
VII. Various Surface Reactions Reported by Other Workers   63
References   69

4/ ADVANCED PROBLEMS IN LIQUID METAL CATALYSIS   74

I. Introduction   74
II. Kinetics   75
III. Stereochemistry   87
IV. Problems in Catalysis by Liquid Alloy   94
V. Electronic Aspect of Catalysis   111
References   122

5/ SURFACE TENSION OF LIQUID METALS AND ALLOYS   125

I. Introduction   125
II. Various Problems Related to Surface Tension   126
References   150

6/ OPTICAL PROPERTIES AND SURFACE TRANSITION ZONE   155

I. Introduction   155
II. Methods of Studying Optical Properties of Liquid Metals   156
III. Overview of Experimental Results   163
IV. Surface Transition Zone   168
References   177

7/ ELECTRON SPECTROSCOPIES AND RELATED SUBJECTS   180

I. Introduction   180
II. Electronic Structures   181
III. Surface Composition   193
References   199

AUTHOR INDEX   203
SUBJECT INDEX   211

Catalysis and Surface Properties of Liquid Metals and Alloys

1
History and Scope

I.	Introduction	1
II.	Historical Survey	2
III.	Motive for Studying the Catalysis of Liquid Metal	4
IV.	Scope of Catalyst Elements	6
	References	6

I. INTRODUCTION

In this chapter important earlier research related to liquid metal catalysis and its scientific background will be described first, in order to clarify the roles played by this work in advancing the study of heterogeneous catalysis. Then the author's motive for initiating and promoting systematic studies on liquid metal catalysis will be described. The purpose of this chapter is to describe the scientific foresight of earlier researchers and to evaluate the position of the work of the author and his co-workers in the history of the catalysis research.

II. HISTORICAL SURVEY

This survey begins with Ipatiew's work which was published in the beginning of this century (1901) [1]. He found that metallic zinc catalyzed the decomposition of alcohols, even above its melting point. Although metallic zinc does not always liquefy at the melting point when it is covered by an oxide layer, Ipatiew's finding is sufficient to suggest that the catalyst metal might possess a catalytic activity, even in its liquefied state. However, Ipatiew gave no special attention to the catalytic activity of liquid metal, and thus 20 years or more elapsed without any significant advances in the research on the catalysis of liquid metal.

In 1925 Taylor [2] published the concept of active centers, which greatly stimulated catalysis researchers and brought about intensive controversy. Several workers postulated that the active centers, areas with special structures, would disappear on fusion and therefore the catalytic activity of a solid metal would disappear or change drastically at the melting point. Thus, they measured the activities of metals below and above the melting points. For instance, Hartman and Brown [3] found that the nitrobenzene hydrogenating activity of a supported cadmium (Cd) catalyst exhibited a maximum at the melting point of Cd (320.9°C) (Fig. 1a), which supported Taylor's opinion. Steacie and Elkin [4] argued against this and considered that Hartman's data merely indicated a reduction in the degree of dispersion of Cd particles on the surface of the catalyst support. Steacie asserted that a pure metal, instead of a supported metal, had to be used as a catalyst to investigate the validity of the concept of active centers. Thus, he employed metallic zinc as a catalyst for the methanol decomposition reaction and found that no discontinuous changes in the catalytic activity took place at the melting point of zinc (419.5°C) (Fig. 1b). However, Adadurow and Didenko [5] suspected that the catalytic activity observed by

Historical Survey

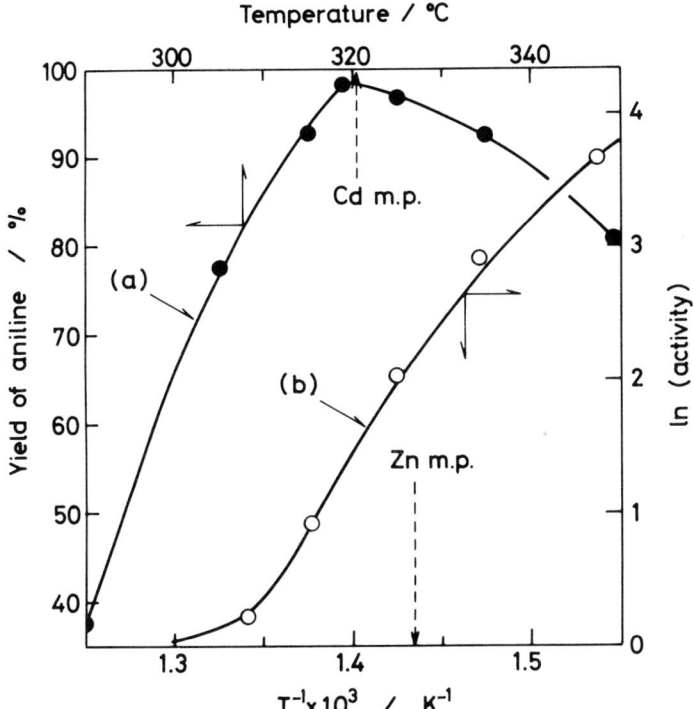

Fig. 1 Temperature dependence of catalytic activities. (a) Nitrobenzene hydrogenating activity of a supported Cd catalyst [3]; (b) methanol decomposition activity of a metallic zinc catalyst [4].

Steacie might be the activity of the oxide layer covering the surface of metallic zinc. Schwab and Martin [6] also argued against Steacie's result and showed that pure zinc liquid was inactive and that only the liquid zinc contaminated by oxygen was active.

Although Weller and his co-workers [7] examined liquid tin Sn(liq.) as a coal liquefaction catalyst, little work on the catalysis of liquid metal had been done until 1960 when Schwab [8,9] published his work on catalysis by liquid alloys. He measured the rate of decomposition of formic acid over two series of liquid alloys (Hg-based binary alloys and Tl-based binary alloys) and interpreted the experimental results on the basis of the electron theory of metals. Unfortunately, however, information about the electronic properties of liquid metals and liquid alloys was not sufficient for his purpose, and he had to use the electron theory of alloys in the solid state. In addition, Schwab confined himself to the study of formic acid decomposition. Therefore, various aspects characterizing the catalysis of liquid metals and liquid alloys remained to be discovered.

To the author's knowledge, no industrial use of the liquid metal catalyst has been realized, although a few patents [10] have claimed the use of a liquid metal catalyst in petroleum decomposition. In this connection, Meszaros's report [11] is cited here. He reported the use of a "melt bed reactor" in producing furan from furfural:

$$PbO + \text{(furyl)}-CHO \longrightarrow \text{(furyl)}-COOH + Pb(liq.) \qquad (1)$$
$$\downarrow$$
$$\text{(furyl)} + CO_2$$

$$Pb(liq.) + 1/2\ O_2 \longrightarrow PbO \qquad (2)$$

In the upper half of this reaction, furfural reacts with PbO and produces furan and Pb(liq.), which goes down toward the bottom of the reactor. In the lower part of the reactor, the Pb(liq.) is oxidized by air and rises due to buoyancy toward the reaction zone. Clearly the reaction of lead is cyclic but not catalytic.

III. MOTIVE FOR STUDYING THE CATALYSIS OF LIQUID METAL

It is easy to suppose that studies of the catalysis of liquid metals would have the following advantages.

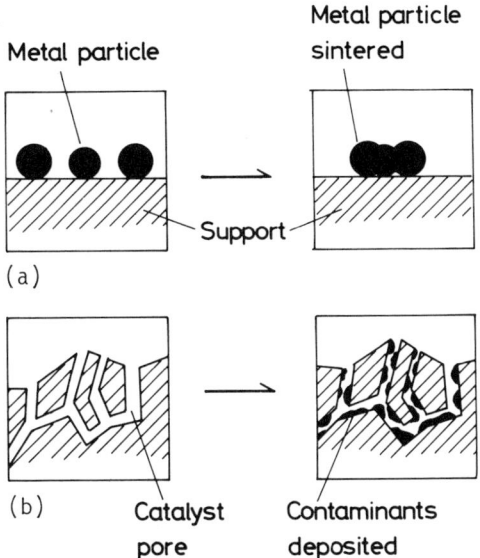

Fig. 2 Models for typical deactivation processes of catalysts: (a) sintering, (b) fouling.

1. No sophisticated techniques are required in preparing the catalyst, although precautions against impurities and contaminations of the catalyst metal are necessary.
2. The deactivation due to sintering (Fig. 2) never takes place.
3. Carbonaceous materials formed by side-reactions are automatically separated from the bulk of the liquid metal due to buoyancy and therefore the deactivation due to the catalyst fouling can be avoided by the use of an appropriate reactor; a catalyst fouling as illustrated in Fig. 2 would never take place.

Thus, the experimental data will be reproducible and universal, and they will provide a firm basis to the scientific discussion on surface catalysis.

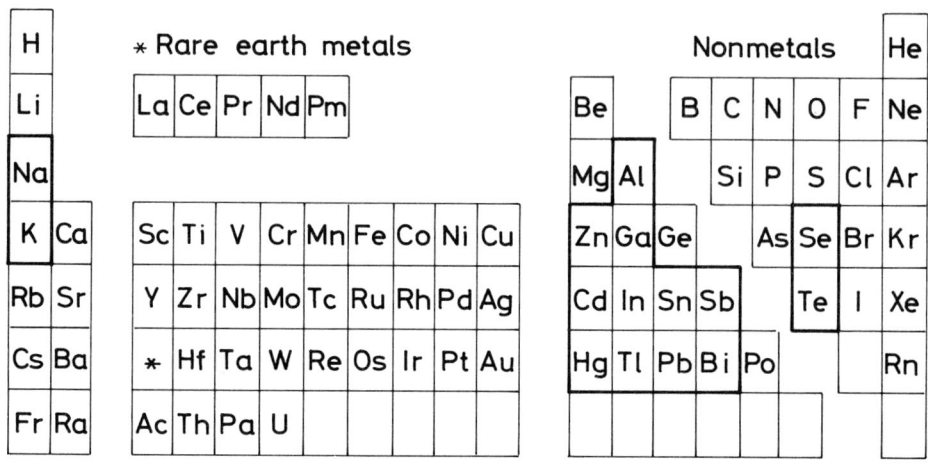

Fig. 3 Elements examined as catalysts in their molten states.

IV. SCOPE OF CATALYST ELEMENTS

Recognizing the advantages mentioned in the preceding section, the author and his co-workers had intended to carry out a series of studies regarding the catalysis of liquid metals and liquid alloys. A previous review [12] summarized somewhat earlier work. The contents of this book include those of the previous review in more detail together with recent results. Outlined by bold frames in Fig. 3 are the elements examined in the catalysis studies carried out over the past 15 years (from 1970 to 1985).

REFERENCES

1. W. Ipatiew, *Ber.*, **34**, 3579 (1901).
2. H. S. Taylor, *Proc. Roy. Soc. London*, **A108**, 107 (1925).
3. R. J. Hartman and O. W. Brown, *J. Phys. Chem.*, **34**, 2651 (1930).
4. E. W. R. Steacie and E. M. Elkin, *Proc. Roy. Soc. London*, **A142**, 457 (1933).

References

5. I. E. Adadurow and P. D. Didenko, *J. Amer. Chem. Soc.*, *57*, 2718 (1935).
6. G. M. Schwab and H. H. Martin, *Z. Elektrochem.*, *43*, 610 (1937).
7. S. Weller, M. G. Pelipetz, S. Friedman, and H. H. Storch, *Ind. Eng. Chem.*, *42*, 330 (1950).
8. G. M. Schwab, *DECHEMA Monogr.*, *38*, 205 (1960).
9. G. M. Schwab, *Ber. Bunsenges. Physik. Chem.*, *80*, 746 (1976).
10. E.P. 221,559 (1923); E.P. 278,235 (1927); U.S. Patent 1,602,310 (1923); U.S. Patent 1,672,459 (1925).
11. L. Meszaros, *Complete Catalytic Laboratories*, Jozsef Attila University, Metrimpex-Labor, Szeged-Budapest, 1963, p. 47.
12. Y. Ogino, *Catal. Rev.-Sci. Eng.*, *23*, 505 (1981).

2
Techniques for Measuring Catalytic Activity

I.	Introduction	8
II.	Catalyst Purification	9
III.	Various Reactors	10
	A. Bubbling-Type Reactor	10
	B. Rectangular Duct Reactor	12
	C. Low-Pressure Reactor	15
	D. Pulse Reactor	16
	E. Reactor with Streaming Catalyst Vapor	17
	F. High-Pressure Autoclave	20
	G. Reaction Techniques in Literature	22
	References	23

I. INTRODUCTION

Activity measurement is one of the most fundamental experiments in catalysis research and the methodology for measuring activities of solid state catalysts is almost completely established. In contrast to this, the activity measurement itself has been a problem to be solved by a number of trials and experiences in the research of the liquid metal catalysis. The problem has only partly been solved and hence the situation mentioned above

is still continuing. Thus, most of the experimental methods
that will be described in this chapter will probably need further refinement.

II. CATALYST PURIFICATION

Although most of the catalyst metals with high purity are commercially obtainable, purification using the technique illustrated in Fig. 1 [1] is recommended, even in an ordinary activity test. The raw metal, charged in a catalyst preparatory
tube, is heated well beyond its melting point and then purified
hydrogen is passed through. The oxide of the catalyst metal is
either reduced or separated, together with other contaminants,
from the bulk of the liquid metal by buoyancy. Then the direction of the hydrogen stream is reversed, causing a transfer of
the liquid metal from the catalyst preparatory tube to the reactor. The hydrogen pressure is so controlled that only a pure
portion of the liquid metal can be taken into the reactor.
Finally, the pipeline connecting the preparatory tube and the
reactor is sealed by fusion. The preparation of a liquid alloy
[2] is carried out in a manner similar to that mentioned above.
For the catalyst purification mentioned above, at least a hydrogen purification device is necessary. Additional connections

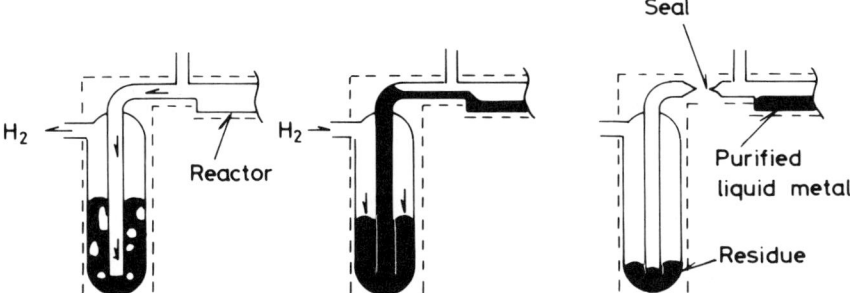

Fig. 1 Procedures for catalyst purification.

to an appropriate vacuum line and to any inert gas supply lines with the gas purifiers are recommended for a rapid and smooth catalyst purification.

III. VARIOUS REACTORS

A. Bubbling-Type Reactor [3-6]

A conventional flow-type catalytic reaction system equipped with a bubbling reactor is illustrated in Fig. 2. This apparatus is quite convenient for a rapid determination of the catalytic activity of a liquid metal or a liquid alloy. A desired reactant supplied from a microfeeder is evaporated in a preheater and then forced to bubble into the catalyst liquid from a port located at the bottom of the reactor. The reactant vapor can always contact with pure liquid metal because impurities such as oxides and carbonaceous materials formed by side-reac-

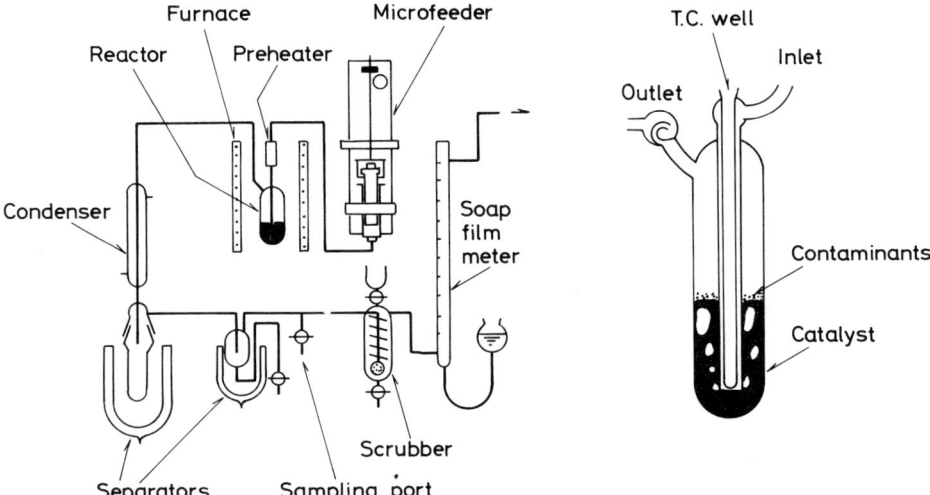

Fig. 2 Flow-type catalytic apparatus equipped with a bubbling-type reactor. (The details of the reactor are shown on the right-hand side of the figure [3-6].)

Various Reactors   11

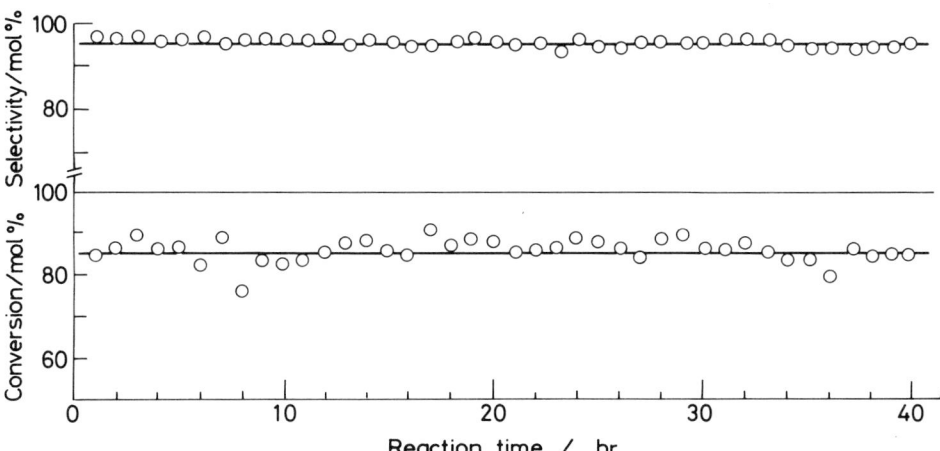

Fig. 3 The cyclohexanol dehydrogenation activity of a liquid zinc catalyst as a function of the reaction time at 500°C [7].

tions are separated automatically from the bulk of the catalyst by buoyancy. Thus, little depression in the catalytic activity by the fouling takes place as illustrated in Fig. 3 [7].

Various types of bubbling reactor are schematically shown in Fig. 4a-f [7]. Except for (f), which is made from stainless steel, the reactors are made from Pyrex glass. Although type (a) is most favorable, any one of (b-d) serves for an ordinary activity test. Usually the stirring of the liquid metal catalyst is made by the bubbling of the reactant vapor, and hence no special stirring device is needed. Thus, the reactor of type (e) is used only for special cases, e.g., studies of the effects of the stirring rate upon the conversion of the reactant. The metallic reactor (f) is not convenient for a catalytic study under ordinary pressure or under low pressure: Special connectors and seals between glass parts and metal parts are needed and, in addition, opening and closing of the reactor are tedious.

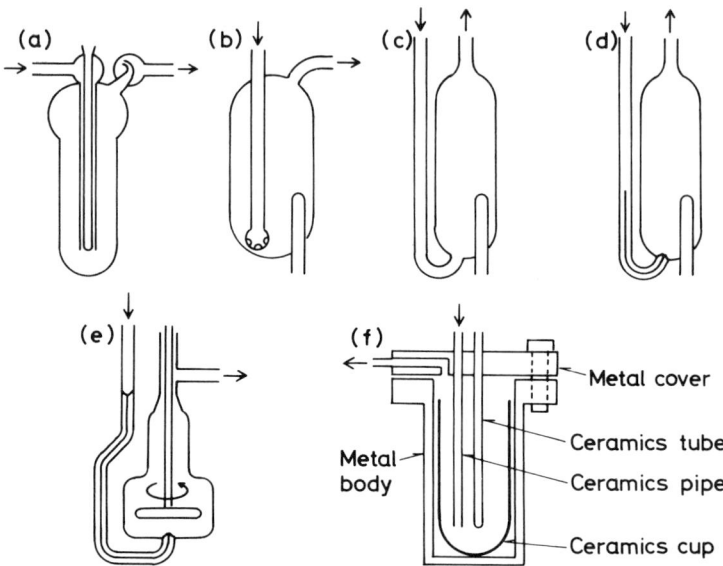

Fig. 4 Several different bubbling-type reactors [7].

B. Rectangular Duct Reactor [8,9]

A schematic drawing of the rectangular duct reactor is given in Fig. 5. A flow-type apparatus equipped with this reactor is convenient for the study of the kinetics of reactions catalyzed by the liquid metal.

The kinetic study with this reactor can be carried out within the framework proposed by Solbrig and Gidaspow [8] and simplified by Miyamoto and Ogino [9]. The Cartesian coordinates (η,ξ,ζ) are defined along the three edges of the duct, and by employing a differential grid model, the differential equation representing the concentration gradient of the reactant in the reactor can be integrated numerically with appropriate boundary conditions. For instance, if the reaction is first-order with respect to the reactant, the integration gives the chart shown in Fig. 6. The abscissa Y of this figure is the dimensionless reactor length defined by

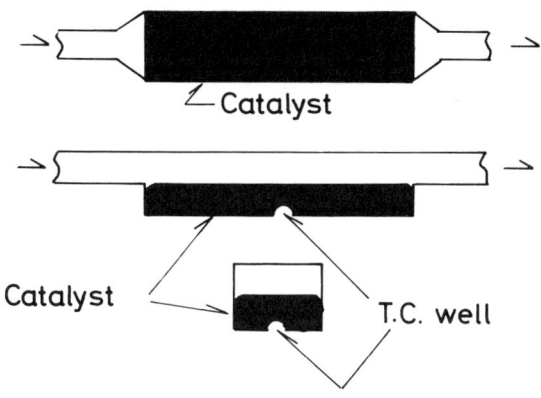

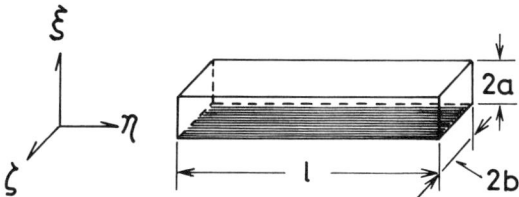

Fig. 5 A rectangular duct reactor containing a liquid metal catalyst [9].

$$Y = \left(\frac{2}{3}\right)\left(\frac{D_A}{Ub}\right)\left(\frac{\eta}{b}\right) \tag{1}$$

where D_A is the effective diffusion coefficient of the reactant and U is the mean linear flow rate of the fluid in the duct. The ordinate C_b^A is the mixing cup concentration of the reactant at an arbitrary position along the η axis. Thus, we can determine the rate constant k' using Fig. 6; the value of Y at $\eta = \ell$ can be obtained from the geometry of the reactor and from the experimental conditions, and the value of the ordinate is obtained from the mixing cup concentration C_b^A measured at the exit end ($\eta = \ell$) of the reactor.

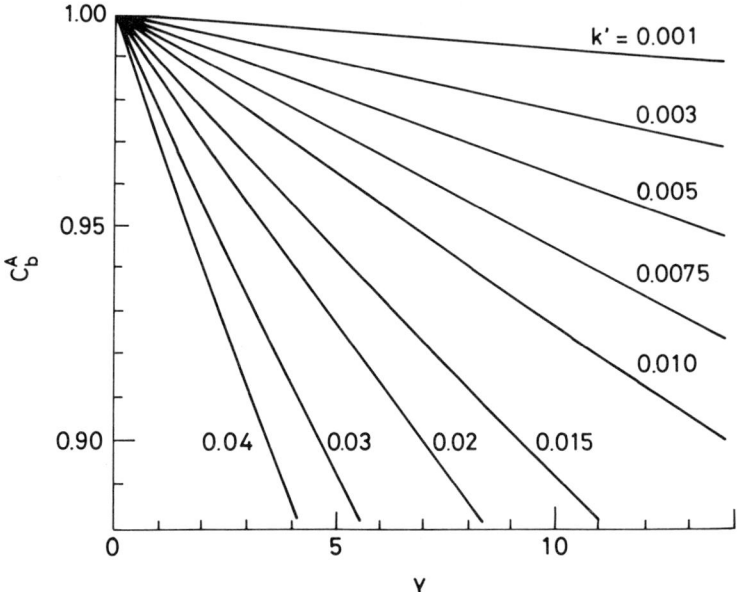

Fig. 6 The mixing-cup concentration c_b^A as a function of the dimensionless reactor length Y [9].

The use of the rectangular duct reactor brings about further merits. Miyamoto and Ogino [9] have proved that the following simple relation holds under appropriate experimental conditions:

$$\left(\frac{1}{k_1^0}\right)(1 + K_A \bar{p}_A) = \frac{(A/SU)}{\ln[1/(1-X)]} \quad (2)$$

where A is the geometrical surface area of the catalyst; S is the cross-sectional area of the duct; X is the conversion defined by $1 - c_b^A$; $\bar{p}_A$ is the mean partial pressure of the reactant vapor streaming in the reactor; and k_1^0 and K_A are the rate constant and the adsorption equilibrium constant, respectively, and both are defined by the following Langmuir-type rate equation, i.e., rate = $k_1^0 c_m^A/(1 + K_A \bar{p}_A)$ where c_m^A denotes the mass concentration of the reactant. Equation 2 indicates that the rate constant k_1^0 and the adsorption equilibrium constant K_A can be obtained simply by measuring the mixing cup concentration C_b^A as a function of the partial pressure ($\bar{p}_A$) of the reactant. It must be noted that great care against the contamination of the catalyst surface is necessary during the measurement of C_b^A.

C. Low Pressure Reactor [1]

The apparatus illustrated in Fig. 7 is available for a kinetic study under low pressure. The reactant vapor circulating through the reaction system contacts with the surface of the liquid metal and reacts. The pressure change caused by the reaction is converted to the change in the resistance of a platinum filament stretched vertically in the mercury manometer and is registered

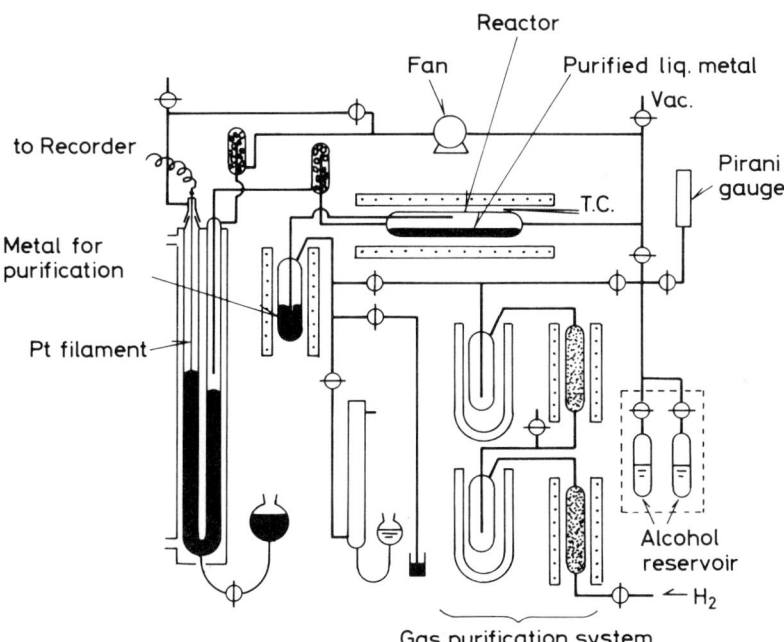

Fig. 7 An apparatus used for kinetic studies over a liquid metal catalyst at low pressures [1].

on a recorder as a function of the reaction time. Thus, we can obtain the relation between the initial rate r_0 and the initial pressure p_A^0 of the reactant vapor.

D. Pulse Reactor [10]

When the reactant is a precious material such as a deutero-compound, the use of a pulse reactor is desirable. Illustrated in

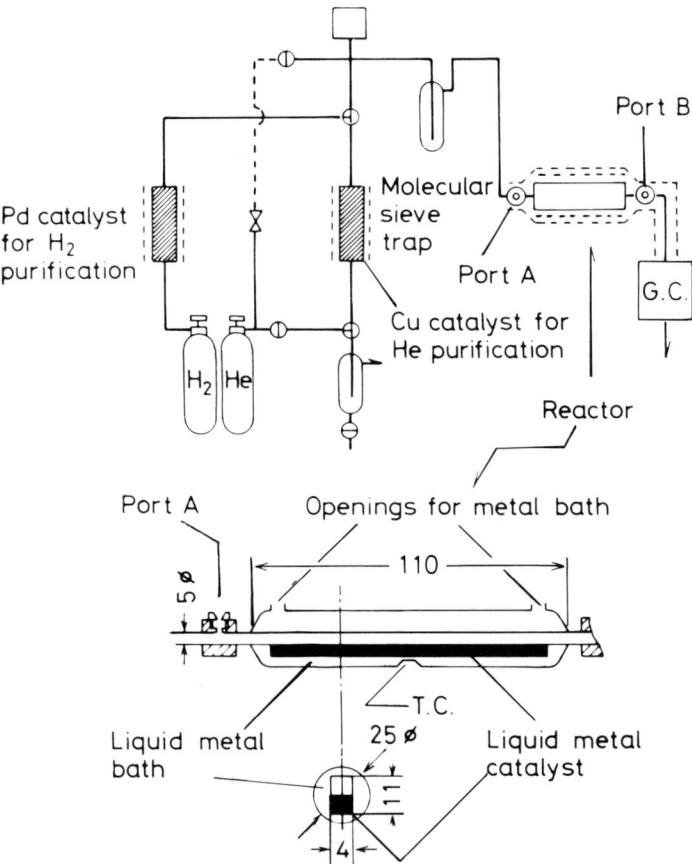

Fig. 8 Pulse reactor containing a liquid metal catalyst (the catalytic zone of the reactor is surrounded by a metal bath to keep a uniform temperature) [10].

Fig. 8 is a pulse reactor that has been applied to the study of
the liquid metal catalysis [10]. It is essentially a modified
rectangular duct reactor that is directly connected to a gas
chromatographic apparatus. As we can see in the figure, there
are two sample injection ports: One is located nearest to the
inlet and the other is located nearest to the outlet of the reactor. The former port serves as an injection port in an ordinary experiment and the latter port serves as an injection port
for measuring the reactor characteristics.

For several typical kinetic cases, the respective relations
between the reactant conversion X and the initial pulse size N_p^0
have been derived theoretically and the results are shown schematically in Fig. 9 [10]. Among four cases shown in the figure,
the one (lower left) in which the surface catalysis obeys the
Langmuir-type kinetics is particularly important. In this case
the following relation has been proved to hold:

$$X = \frac{(2a/A)k_1^0(\ell/u)}{(1 + K_A \alpha N_p^0)} \quad (3)$$

where α is a proportionality constant defined by $p_A^0 = \alpha N_p^0$. This
equation tells us that the rate constant k_1^0, as well as the
adsorption equilibrium constant K_A, can easily be obtained by
measuring the conversion X as a function of the pulse size N_p^0.

E. Reactor with Streaming Catalyst Vapor [11]

Generally, the overall reaction that would take place in a volume element dV of the rectangular duct reactor with dimensions
shown in Fig. 10a is given by

$$\left(\alpha r_w \rho_w + r_t + r_v \rho_v + \frac{r_c \rho_c}{b} \right) dV = F\, dX \quad (4)$$

where α is a geometrical factor given by $(2/a + 1/b)$; r is the
reaction rate; ρ is the density; and the subscripts w, t, v,
and c denote the reactor wall, thermal reaction, catalyst vapor,

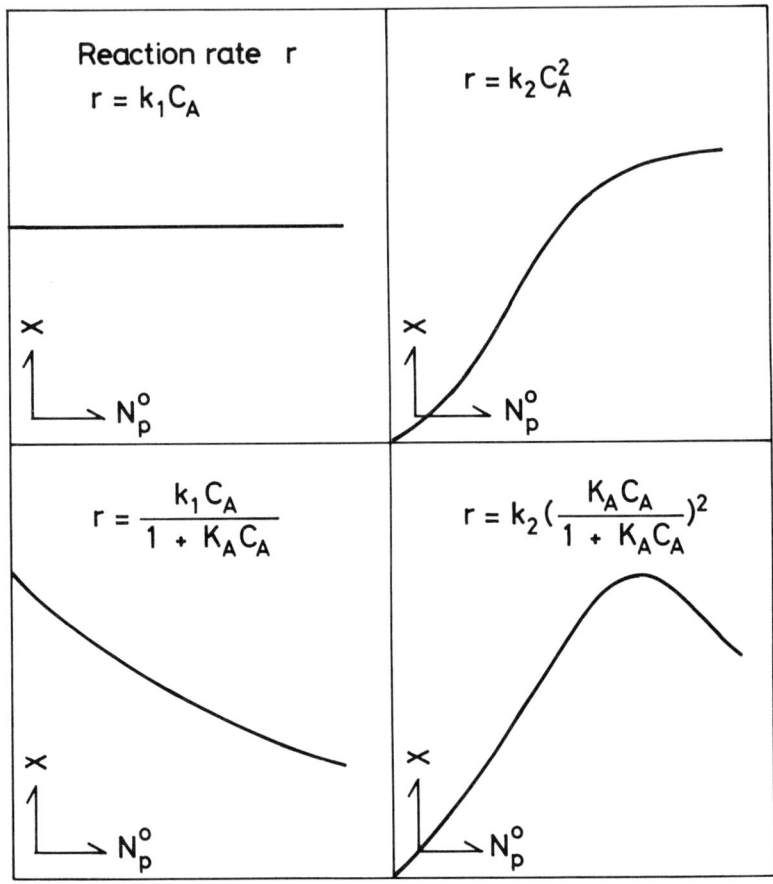

Fig. 9 Schematic criteria for identifying kinetics of pulse reactions over the liquid metal catalyst, where X is conversion; N_p^o, pulse size; k_1, first-order rate constant; k_2, second-order rate constant; K_A, adsorption equilibrium constant; and C_A, concentration of reactant vapor [10].

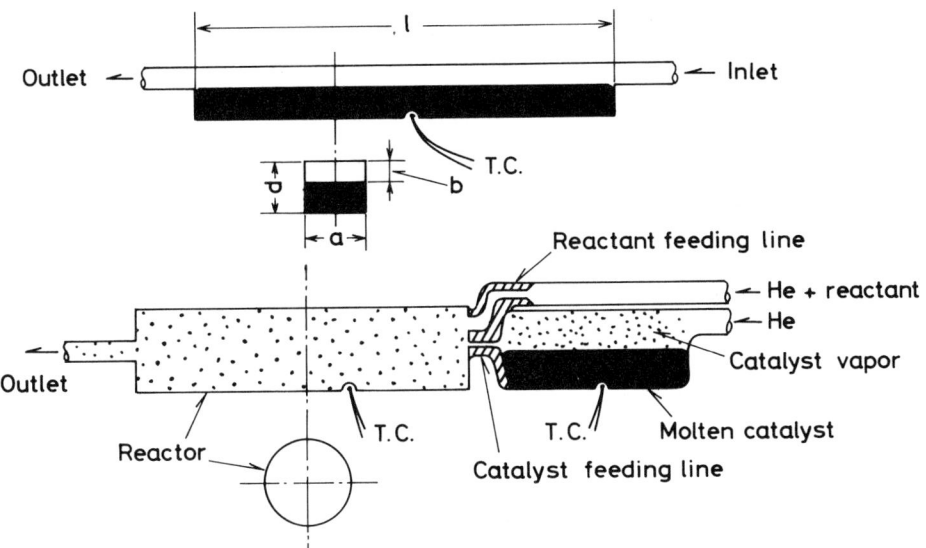

Fig. 10 (a) Rectangular duct reactor containing a liquid metal catalyst having a high vapor pressure. (b) Horizontally placed cylindrical reactor with a streaming catalyst vapor [11].

and the liquid metal catalyst, respectively; F is the feed rate of the reactant; and X is the conversion.

When the reaction at the inner wall of the reactor is insignificant, we can assume that $r_w = 0$. On the other hand, certain liquid metals such as Zn(liq.), Cd(liq.), Na(liq.), K(liq.), and a semimetal Te(liq.) have significantly high vapor pressures at the high temperatures where the activities are measured. In addition, certain reactants undergo thermal reaction. Thus, we have to measure r_t and r_v in order to obtain an intrinsic catalytic activity r_c.

The reactor shown in Fig. 10b serves for the measurements of r_t and r_v. The reactor is essentially a vacant cylinder placed horizontally. The inlet of the reactor is connected to two lines, a catalyst vapor feeding line and a reactant vapor feeding line. A stream of helium saturated with the catalyst vapor is fed through the former line, and another stream of helium containing the reactant vapor is fed through the latter line. At the inlet of the reactor, the two streams join and their components are mixed with each other. The catalysis and pyrolysis take place in the bulk of the mixture flowing through the reactor toward the outlet. The material balanced equation corresponding to Eq. 4 is

$$(r_t + r_v \rho_v) dV = F \, dX \tag{5}$$

Furthermore, if we perform an experiment without the catalyst vapor, the following equation can be used for the analysis of the experimental result:

$$r_t \, dV = F \, dX \tag{6}$$

Thus we can separately determine r_t, r_v, and hence r_c.

The residence time of the mixture in the reactor can be varied by changing either the reactor length or the flow rate of the reactant vapor. The effect of the reactor wall can be checked by changing the radius of the reactor and observing the resulting effect upon conversion (the geometrical factor α is $2/R$, where R is the radius of the reactor). In addition, the outlet line of the reactor has to be cooled well below the temperature at which the reaction virtually ceases to occur.

F. High-Pressure Autoclave [12]

The reactor illustrated in Fig. 11 is an autoclave used for a coal liquefaction over a liquid metal catalyst. This autoclave has been designed under the following precautions.

1. An effective stirring during the reaction is necessary in order to make intimate contact between the catalyst and the reacting coal. In addition, the use of metallic stirrer is not favorable because certain liquid metals attack it.

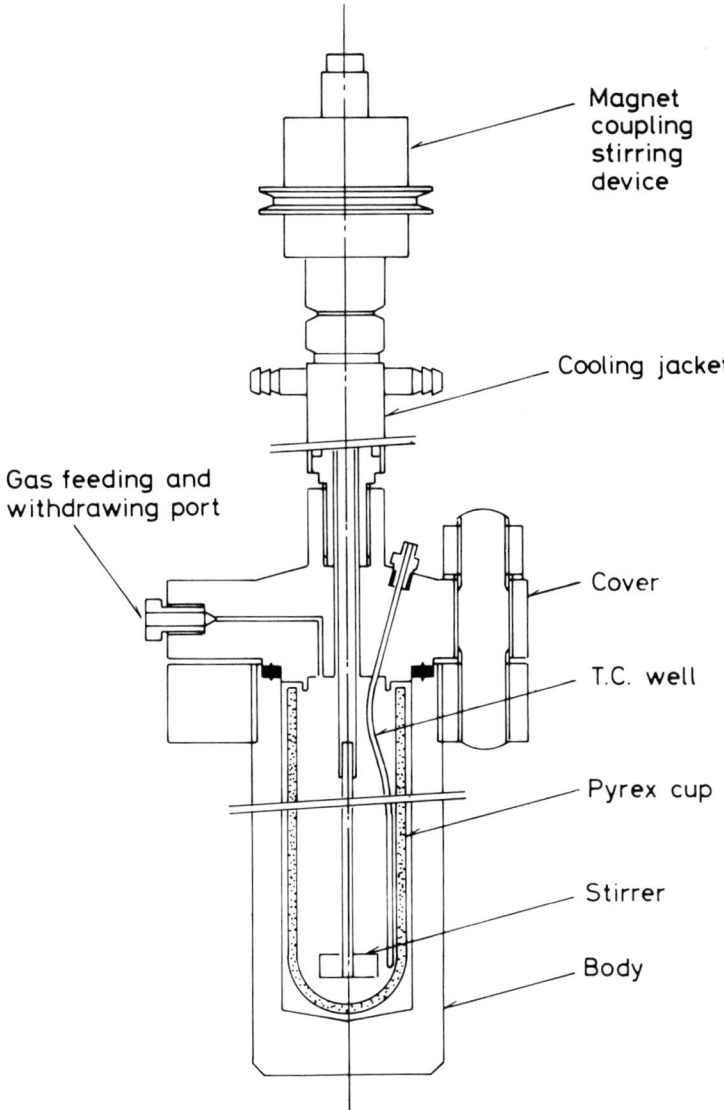

Fig. 11 Autoclave used for coal liquefaction over the liquid metal catalyst [12].

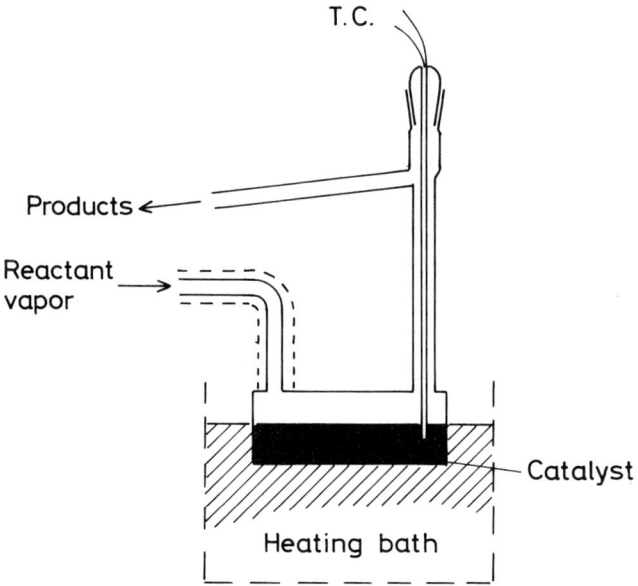

Fig. 12 Reactor design proposed by Schwab [13].

2. The corrosion of the inner wall of the autoclave by the liquid metal has to be minimized, and, in addition, handling of the reaction mixture before and after the activity measurement should be easy.

By using a glass stirrer and by placing a glass cup reactor in the autoclave, the requirements mentioned above have been satisfied quite well.

G. Reaction Techniques in Literature

Schwab [13] proposed the use of a reactor as shown in Fig. 12. The main purpose of this design was to facilitate an optical observation of contaminations of the catalyst surface. A second aim was to minimize the ratio of the glass surface area exposed to the reactant vapor at a high temperature to the catalyst surface area. According to the author's experience, the use of an electric furnace with two vertical slit-like windows mutually locating on the opposite sides of the wall of the furnace enable us to make visual and optical inspections of the catalyst surface.

Addison [14] and his co-workers made intensive studies on liquid alkali metals and developed many experimental techniques. Although they did not study the surface catalysis, a unique and simple magnetic stirring device and an electromagnetic pump would probably contribute to improving the catalytic reaction systems.

REFERENCES

1. Y. Saito, F. Miyashita, and Y. Ogino, *J. Catal.*, *36*, 67 (1975).
2. Y. Saito, H. Yoshida, T. Yokoyama, and Y. Ogino, *J. Colloid Interf. Sci.*, *66*, 440 (1978).
3. Y. Saito, A. Miyamoto, and Y. Ogino, *Kogyo Kagaku Zassi*, *74*, 1521 (1971).
4. K. Kashiwadate, Y. Saito, A. Miyamoto, and Y. Ogino, *Bull. Chem. Soc. Jpn.*, *44*, 3004 (1971).
5. Y. Saito, N. Hiramatsu, N. Kawanami, and Y. Ogino, *Bull. Jpn. Petrol. Inst.*, *14*, 169 (1972).
6. Y. Ogino, Y. Saito, and K. Okano, *Semi-annu. Rep. Asahi Glass Found. Contrib. Ind. Technol.*, *22*, 37 (1972).
7. Y. Saito, Ph.D. thesis, Tohoku University, 1978.
8. C. W. Solbrig and D. Gidaspow, *AIChE. J.*, *13*, 346 (1967).
9. A. Miyamoto and Y. Ogino, *J. Catal.*, *27*, 311 (1972).
10. A. Miyamoto and Y. Ogino, *J. Catal.*, *37*, 133 (1975).
11. K. Takahasi and Y. Ogino, *Fuel*, *60*, 975 (1981).
12. M. Matsuura, S. Matsunaga, S. Ozawa, and Y. Ogino, *Fuel*, *62*, 407 (1983).
13. G. M. Schwab, *Ber. Bunsenges. Physik. Chem.*, *80*, 746 (1976).
14. C. C. Addison, *The Chemistry of the Liquid Alkali Metals*, John Wiley & Sons, Chichester, New York, Brisbane, Toronto, Singapore, 1984.

3
Reactions over Liquid Metals and Alloys

I.	Introduction	25
II.	Dehydrogenation of Alcohols	26
III.	Dehydrogenation of Amines	33
IV.	Dehydrogenation of Hydrocarbons	36
V.	Hydrogen Transfer Reactions	43
	A. Reactions between Alcohols and Ketones	43
	B. Reactions between 2-Butanol and Unsaturated Aromatic Hydrocarbons	49
VI.	Coal Liquefaction	49
	A. Liquefaction with Sn(liq.)	49
	B. Liquefaction Activities of Other Liquid Metals	58
	C. Model Compound Reaction over Liquid Metal	58
	D. Asphalt Digestion	62
VII.	Various Surface Reactions Reported by Other Workers	63
	A. Research with Classical Techniques	63
	B. Research with Special Techniques	66
	References	69

I. INTRODUCTION

The most important and urgent problem in an early stage of studying a new type of catalyst is to find the scope of possible reactions by activity tests. Studies on microscopic mechanisms of catalysis as well as on structural details of the catalyst are significant only when the catalyst being studied is active.

Under the strategy mentioned above, a number of trials were made [1-6] to find that certain lower alcohols can be dehydrogenated by such liquid metal catalysts as Zn(liq.), Al(liq.), Ga(liq.), In(liq.), and Tl(liq.). In the meantime lower amines have also been found [1,7,8] to be dehydrogenated over the same catalysts as those mentioned above. This finding was somewhat surprising because it had been believed that amines would not be dehydrogenated by liquid metal catalysts [9]; the reason is that any simple amine does not contain oxygen atoms in its molecule and hence it cannot make the catalyst surface oxidized and active.

Extended dehydrogenation studies with terpene alcohols and other higher alcohols, including aromatic alcohols, have shown that the catalysts mentioned above are active for the reactions examined [1,10-12]. Since some of these reactions are important practically, methods of increasing the conversion and selectivity have been studied somewhat in detail. For this purpose, hydrogen transfer reactions between alcohols and ketones have been found to be promising [1,10,11,13]. Furthermore, an interesting stereochemical aspect of the catalysis over the liquid metal has been revealed during the study on the hydrogen transfer reaction [1,14-16].

From the early stages of research, the dehydrogenation of hydrocarbons frequently has been examined. However, this work was unsuccessful until a semimetal, Te(liq.), was found as the catalyst for the dehydrogenation of polynuclear aromatic hydrocarbons [17,18]. In addition, some liquid alloys containing either Na or K have been found to catalyze the dehydrogenation of certain aliphatic hydrocarbons as well as aromatic hydrocarbons [19-21].

The success in hydrocarbon reactions greatly encouraged the author and his co-workers, who have been attempting to examine the liquid metal catalyst for much more complex reactions such as coal liquefaction and digestion of petroleum residues [22-29]. Studies of these subjects have revealed new aspects of the catalysis of liquid metals; important roles played by the interaction between free radicals and the liquid metal in the reaction have been recognized. Thus, studies of complex free radical reactions over liquid metal catalysts are intended, and part of the experimental results for a model compound reaction have been published [30].

Although this chapter serves mostly for describing the chemical reactions catalyzed by liquid metals, descriptions of recent papers reporting chemical reactions related to the surface chemistry of liquid metals have also been included in the last section.

II. DEHYDROGENATION OF ALCOHOLS

Information about the dehydrogenation of alcohols over the liquid metal catalysts is presented in Table 1. These reactions are catalyzed by Al(liq.), Ga(liq.), In(liq.), Tl(liq.), and Zn(liq.). In(liq.) is most favorable among the active metals for use in scientific studies [5,6] because indium metal of the highest purity (99.999%) can be obtained commercially at a reasonable price. It is soft and easy to handle, its melting temperature (156.4°C) is moderate, and its vapor pressure is low. On the other hand, aluminum is apt to form an irreducible and catalytically unfavorable oxide and, in addition, its melting point (660.1°C) is somewhat higher than the temperature where ordinary

TABLE 1 Various Reactant Alcohols and the Main Products

Reactants[a]	Products[a]	Refs.
—OH	=O, O–CH–O–	[1, 2]
CH₃CH₂-OH	CH₃CHO	[1, 4, 10]
CH₃CH₂CH₂-OH	CH₃CH₂CHO	[1, 4]
(CH₃)₂CH-OH	(CH₃)₂C=O	[1, 4, 6]
CH₂=CHCH₂-OH	CH₂=CHCHO, CH₃CH₂CHO	[1, 4]
CH₃(CH₂)₂CH₂-OH	CH₃CH₂CH₂CHO	[1, 3]
(CH₃)₂CHCH₂... -OH (sec-butanol)	CH₃CH₂C(=O)CH₃	[1, 3, 5, 12]
(CH₃)₂CHCH₂-OH	(CH₃)₂CHCHO	[1, 3]
(CH₃)₃C-OH	decomposition	[1]
CH₃(CH₂)₆CH₂-OH	CH₃(CH₂)₆CHO	[1]
3,7-dimethyloctan-1-ol	3,7-dimethyloctanal	[1, 11]
geraniol	citral	[1, 11]
nerol	nerol-aldehyde, decomp.	[1, 11]

TABLE 1 (continued)

Reactants[a]	Products[a]	Refs.
(CH₃)₂C=CH-CH₂-CH₂-CH(CH₃)-CH₂-CH₂-OH	(CH₃)₂C=CH-CH₂-CH₂-CH(CH₃)-CH₂-CHO	[1, 11]
(CH₃)₂C(OH)-CH₂-CH₂-CH₂-CH(CH₃)-CH₂-CH₂-OH	(CH₃)₂C(OH)-CH₂-CH₂-CH₂-CH(CH₃)-CH₂-CHO	[1, 11]
HO-CH₂-CH₂-OH	decomp.	[11]
CH₃-O-CH₂-CH₂-OH	decomp.	[11]
HO-CH₂-CH₂-CH₂-CH₂-OH	tetrahydrofuran ; γ-butyrolactone	[1, 10]
HO-CH(CH₃)-CH(CH₃)-OH	decomp.	[10]
HO-CH₂-CH(CH₃)-CH₂-OH	decomp.	[10]
cyclobutanol	decomp.	[10]
cyclopentanol	cyclopentanone	[1, 10]
cyclohexanol	cyclohexanone	[1, 4, 10]
2-methylcyclohexanol	2-methylcyclohexanone	[1, 10]
cycloheptanol	cycloheptanone	[1, 10]
cyclooctanol	cyclooctanone	[1, 10]
cyclodecanol	cyclodecanone	[1, 10]

TABLE 1 (continued)

Reactants[a]	Products[a]	Refs.
(bornanol, OH)	(camphor, =O)	[1, 11]
(menthol)	(menthone)	[1, 11]
(benzyl alcohol)	(benzaldehyde)	[1, 10]
(2-phenylethanol)	(phenylacetaldehyde)	[1, 10, 11]
(1-phenylethanol)	(acetophenone)	[10]
(4-methoxybenzyl alcohol)	(4-methoxybenzaldehyde)	[1, 10, 11]
(perillyl-type cyclohexanol)	(open-chain aldehyde)	[1]
(myrtenol)	(myrtenal), (OH diene), (OH diene)	[1]

[a]The reactants and products are expressed by formulas composed of the following units: —: $-\overset{|}{\underset{|}{C}}-\overset{|}{\underset{|}{C}}-$; =: $\overset{\diagdown}{\diagup}C=C\overset{\diagup}{\diagdown}$; —OH: $-\overset{|}{\underset{|}{C}}-OH$; =O: $\overset{\diagdown}{\diagup}C=O$; —O—: $-\overset{|}{\underset{|}{C}}-O-\overset{|}{\underset{|}{C}}-$.

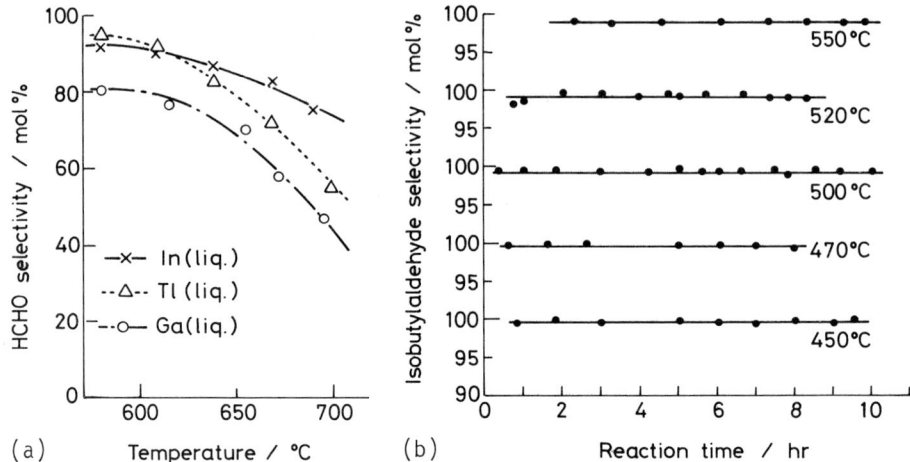

Fig. 1 Experimental data showing selectivities of some liquid metal catalysts in the dehydrogenation of alcohol. (a) The formaldehyde selectivity in methanol dehydrogenation over In(liq.), Tl(liq.), and Ga(liq.); (b) the isobutylaldehyde selectivity in isobutanol dehydrogenation over Zn(liq.) [2].

alcohols are thermally stable. Thus, Al(liq.) shows a lower selectivity in the dehydrogenation of alcohol [2]. Gallium is very expensive and thallium is toxic. Zinc liquid shows a remarkably high activity [1-3], but its high vapor pressure and high susceptibility to oxidation bring considerable confusion into the interpretation of the experimental results, as has been described in the previous chapter (Chap. 1, II).

Most primary alcohols and secondary alcohols are converted to the corresponding aldehydes and ketones over the liquid metal catalyst. Except for a few cases, the selectivities of these reactions are satisfactorily high, as exemplified in Fig. 1 [2]. Fig. 1a shows high selectivities of In(liq.), Tl(liq.), and Ga(liq.) for the methanol dehydrogenation, which is practically important in producing a water-free mixture of formaldehyde and methanol. Fig. 1b represents a very high selectivity of Zn(liq.)

for the dehydrogenation of isobutanol and exemplifies the high selectivities that are frequently observed in the dehydrogenation of other alcohols over this catalyst. Tertiary alcohols are affected little by the liquid metal catalyst and decompose at high temperatures [1].

The reactivities of polyhydric alcohols are still uncertain; although ethylene glycol, 1,3-butandiol, and 2,3-butandiol undergo decomposition [10], 1,4-butandiol is dehydrogenated to form γ-butyrolactone (tetrahydrofuran is the main byproduct)

$$HO-(CH_2)_4-OH \longrightarrow \text{[cyclic]}=O + 2H_2; \text{[cyclic]} \qquad (1)$$

Results for unsaturated alcohols are somewhat complicated [1,4]. Although the corresponding unsaturated aldehydes are formed by the dehydrogenation, saturated aldehydes and decomposition products are found in the products, e.g.,

$$CH_2=CH-CH_2-OH \longrightarrow CH_2=CH-CHO \qquad (2)$$
$$CH_3-CH_2-CHO$$
$$C_3H_6, C_2H_4, C_2H_6, ---$$

and

$$\text{[terpene alcohol]} CH_2OH \longrightarrow \text{[terpene aldehyde]} CHO \qquad (3)$$

The selectivity of dehydrogenation of a particular alcohol varied widely, depending on the catalyst used and the reaction conditions. As we can see later (Section V), the coexistence of acetone or another ketone greatly improves the selectivity of dehydrogenation of an unsaturated terpene alcohol [1].

Cycloalkanol as well as aromatic alcohols are also dehydrogenated with high selectivities over the liquid metal catalyst

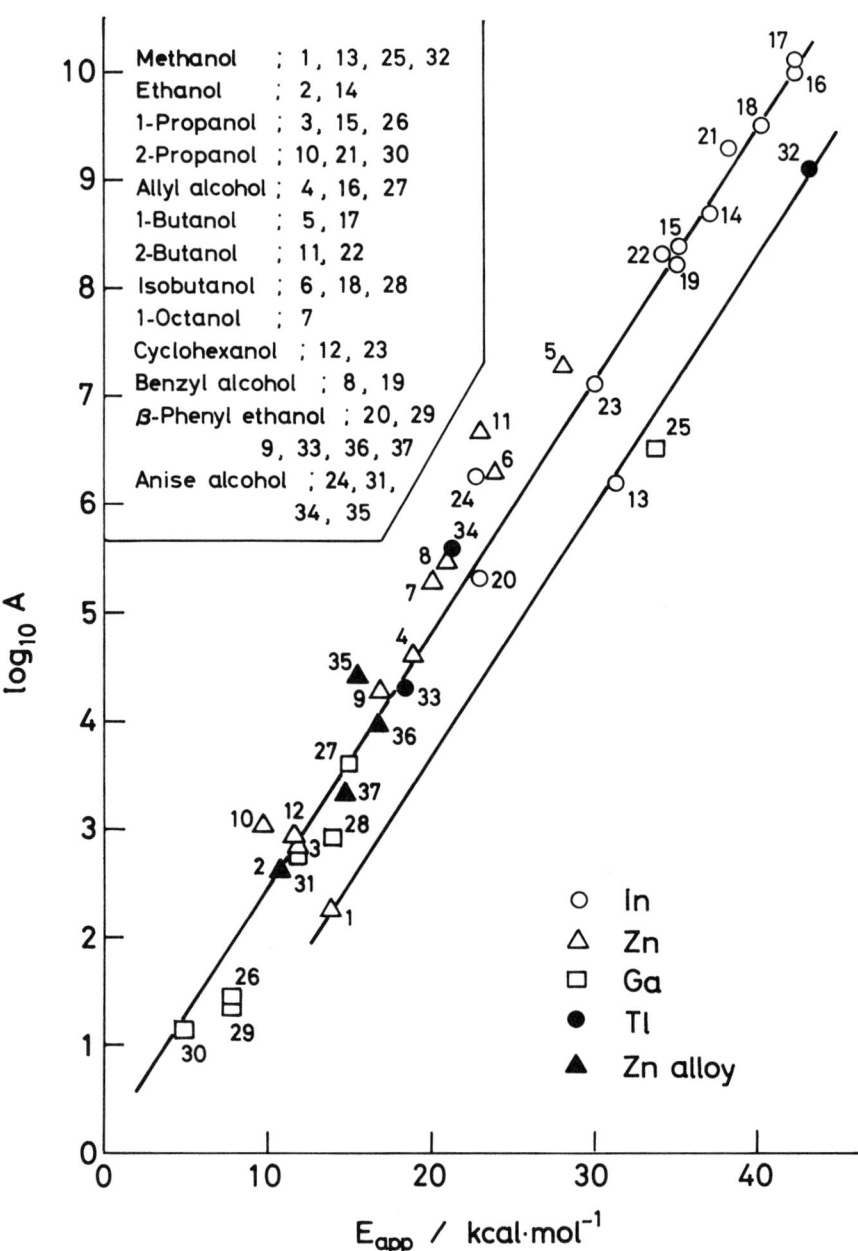

Fig. 2 Summarized plots for alcohol dehydrogenation activities of liquid metal catalysts [1].

[1,10]; when the reactant alcohol is in a viscous or in a solid state, an appropriate solvent has to be used (the most favorable solvent is acetone).

Before closing this section, an interesting relation obtained for the alcohol dehydrogenation is presented in Fig. 2 [1]. The figure is the plots of apparent activation energies E_{app} against the logarithms of the frequency factor A. These kinetic parameters are obtained by assuming an irreversible first-order kinetics. Most experimental data appear to fall on a single straight line, although the data for methanol systematically deviate from this line. This fact appears to suggest that the alcohol dehydrogenation over the liquid metal obeys one simple rule, but it is not yet discovered. An important point is that the figure coupled with Table 1 undoubtedly demonstrates the alcohol-dehydrogenating activities of the liquid metals and liquid alloys mentioned herein.

III. DEHYDROGENATION OF AMINES

Although the activity of Al(liq.) is still uncertain, other liquid metals that are active for the alcohol dehydrogenation are also active for the dehydrogenation of amines [1,7,8]. Among the active liquid metals, Zn(liq.) is active for every amine listed in Table 2, while Ga(liq.) is inactive for the dehydrogenation of benzylamine, cyclohexylamine, and di-n-butylamine [1,7].

Simple normal alkylamines are dehydrogenated to the corresponding nitriles with satisfactorily high selectivities over the Zn(liq.) catalyst, as shown in Fig. 3 [7]. Small amounts of Schiff's base and ammonia are the by-products. On the other hand, the selectivity of dehydrogenation of normal alkylamines over Ga(liq.) is not so high owing to side-reactions forming Schiff's base, ammonia, and dialkylamines corresponding to the reactant amines. The formation of a reactive intermediate,

TABLE 2 Various Reactant Amines and the Main Products

Reactants[a]	Products[a]	Refs.
propylamine	propanenitrile, N-propylidenepropylamine, NH_3, dipropylamine	[1, 7]
allylamine	propynenitrile, propanenitrile, NH_3, propylamine, N-allylidene allylamine	[1, 7]
butylamine	butanenitrile, N-butylidenebutylamine, NH_3, dibutylamine	[1, 7, 8]
sec-butylamine	NH_3, N-(butan-2-ylidene)butan-2-amine	[1, 7]
isobutylamine	isobutanenitrile, N-isobutylideneisobutylamine, NH_3, diisobutylamine	[1, 7]
pentylamine	NH_3, pentanenitrile, N-pentylidenepentylamine, dipentylamine	[1, 7]
dibutylamine	N-butylidenebutylamine	[1, 7]
cyclohexylamine	(N-cyclohexylidenecyclohexylamine)	[1, 7, 8]
benzylamine	benzonitrile	[1, 7, 8]

[a] —N: $-\underset{|}{\overset{|}{C}}-N\diagdown$; =N: $\diagdown C=N-$; ≡N: $-C≡N$.

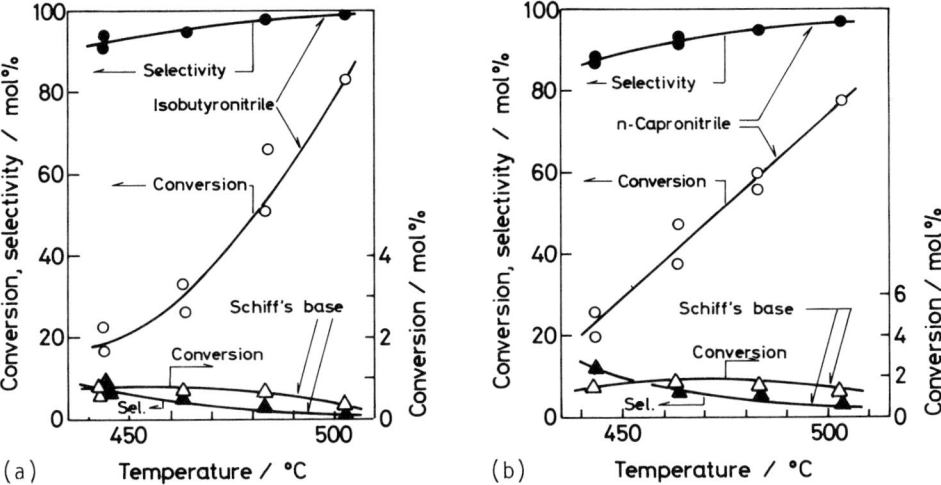

Fig. 3 Activities and selectivities of the Zn(liq.) catalyst.
(a) Dehydrogenation of isobutylamine; (b) dehydrogenation of n-hexylamine [7].

aldimine, appears responsible for the initiation of the side-reactions observed both in the Zn(liq.) and in the Ga(liq.)-catalyzed reactions [1,7,8];

$$R-CH_2-NH_2 \longrightarrow [R-CH=NH] + H_2 \rightarrow R-C\equiv N + 2H_2 \quad (4)$$

$$Zn(liq.), Ga(liq.) \downarrow + R-CH_2-NH_2$$

$$R-CH=N-CH_2R + NH_3 \quad (5)$$

$$Ga(liq.) \downarrow + H_2$$

$$(RCH_2)_2NH \quad (6)$$

The selective dehydrogenation of unsaturated aliphatic amines, e.g., allylamine, have been found to be difficult [7]: The selectivity for acrylonitrile formation is less than 20 mol%.

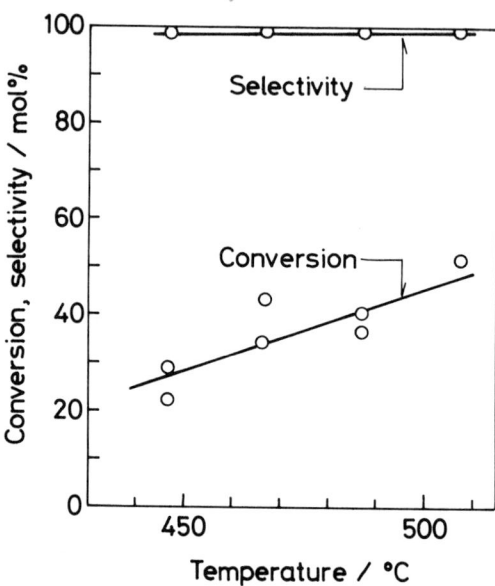

Fig. 4 The high selectivity of benzonitrile formation from benzylamine over the Zn(liq.) catalyst [1,8].

On the other hand, benzylamine, an aromatic amine, has been dehydrogenated with a good selectivity over the Zn(liq.) catalyst as we can see in Fig. 4 [1,8].

IV. DEHYDROGENATION OF HYDROCARBONS

The liquid metals and liquid alloys that have been found to be active catalysts for the dehydrogenation of alcohols are almost inactive for the dehydrogenation of hydrocarbons. Even tetralin, a good hydrogen donor, can not be dehydrogenated by the catalysts mentioned hitherto. However, after a number of trials, a liquid of semimetal, Te(liq.), has been found to catalyze the dehydrogenation of tetralin while it is inactive for the dehydrogenation of alcohols [17]. Furthermore, as we can see in Table 3, Te(liq.) is capable of dehydrogenating various polynuclear hydrocarbons [17,18]. The data shown in the table also show that the catalytic activity of Te(liq.) is greatly enhanced by the incorporation of selenium. The effect of a stoichiometric reaction between hydrocarbon and selenium

$$Se + -CH_2-CH_2- \rightarrow H_2Se + -CH=CH- \qquad (7)$$

has been proved to be trivial, and the activity of the Se-promoted catalyst persists for a long period, as we can see in Fig. 5. It appears worth warning here that appropriate absorbers containing a reagent capable of decomposing the toxic H_2Se vapor have to be attached to the exit line of the apparatus for measuring the catalytic activity of the Te-Se liquid mixture.

Despite the excellent activity of the liquid Te-Se system for the dehydrogenation of polynuclear hydrocarbons, this system is inactive for the dehydrogenation of other hydrocarbons, e.g., ethylbenzene. We have to use much more active catalysts, i.e., catalysts containing alkali metals, to dehydrogenate hydrocarbons other than polynuclear hydrocarbons.

Binary liquid alloys containing either Na or K as an active component are active catalysts for the dehydrogenation of ethylbenzene, as we can see in Fig. 6a, b [19,20]. The activity of the K-containing catalyst is superior to that of the Na-containing catalyst: An addition of 5 atom% of K into any diluent meta metal is sufficient to bring about an activity that is comparable to that of the corresponding liquid alloy containing ~30 atom% of Na. Other alkylbenzenes such as cumene and p-cymene are also dehydrogenated over the liquid alloys mentioned above, as we can see in Fig. 7 [19]. It must be noted that the alloy catalysts are capable of catalyzing the dehydrogenation of polynuclear hydrocarbons [19].

Another important catalysis that takes place over the liquid alloy containing Na or K is a dehydrogenation of butene to form 1,3-butadiene [21]. Any one of the three isomers, as well as the mixture of the isomers, is available as the reactant.

TABLE 3 Catalytic Activities of Te(liq.) and Te-Se(liq.) for the Dehydrogenation of Polynuclear Hydrocarbons [17,18].

Catalysts	Reactant (mol/hr)	Solvent (mol/hr)	Temperature (°C)	Total conversion (%)	Selectivity (%) Individual product		Total
Te(liq.) (70 g)	(0.039)	—	593	15.4	24	72	96
	(0.041)	—	578	9.0	36	39	75
	(0.0033)	Benzene (0.06)	574	9.4	90		90
	(0.031)	Benzene (0.06)	571	3.9	88		88
	(0.0027)	Benzene (0.06)	586	20.4	99		99
	(0.0025)	Benzene (0.05)	581	17.8	79		79
			588	18.2	70	13	83

Dehydrogenation of Hydrocarbons

TABLE 3 (continued)

Catalysts	Reactant (mol/hr)	Solvent (mol/hr)	Temperature (°C)	Total conversion (%)	Selectivity (%) Individual product			Total
	(0.059)		580	35.7	⌬⌬ 11		85	96
			565	24.4	19	⌬⌬	75	94
			550	12.6	34		58	92
			535	9.0	48		40	88
			515	5.9	54		26	80
	(0.0033)	Benzene (0.06)	580	64.6				96
			568	50.1				96
			550	29.5	b			96
			535	19.2				97
			500	11.1				97
Te-Se[a] (liq.)	(0.0027)	Benzene (0.05)	580	96.7				99
			570	92.7				99
			555	81.8	c			99
			540	64.4				98
			520	41.4				95
			500	27.1				93

TABLE 3 (continued)

Catalysts	Reactant (mol/hr)	Solvent (mol/hr)	Temperature (°C)	Total conversion (%)	Selectivity Individual product		Total
	[structure] (0.0025)	Benzene (0.05)	580	92.4	22	72	94
			560	83.1	41	49	90
			545	66.3	49	39	88
			524	45.0	51	30	81
			505	34.0	40	26	66
	[structure] (0.0031)	Benzene (0.06)	580	95.0			90
			560	92.3			93
			540	77.6	d		92
			520	60.9			95
			500	45.1			96

[a] Se: 10 atom%.
[b] Acenaphthylene.
[c] Phenanthrene.
[d] Anthracene.

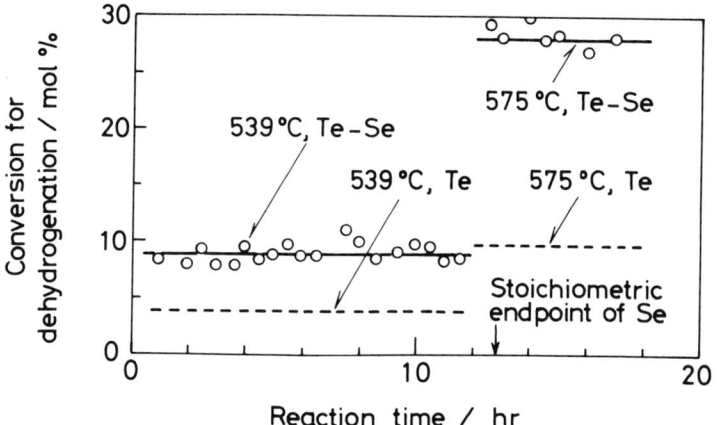

Fig. 5 A strong promoting effect of Se (10 atom%) upon the catalytic activity of Te(liq.) for the dehydrogenation of tetralin [18].

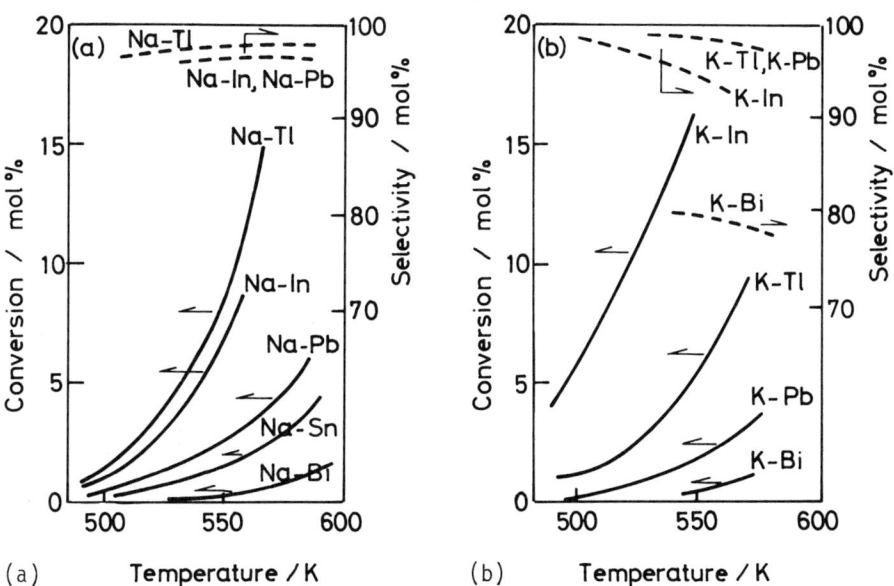

Fig. 6 Ethylbenzene dehydrogenating activities of some binary liquid alloys. (a) Alloys containing Na; (b) alloys containing K [19,20].

TABLE 4 Butene Dehydrogenating Activities of Several Liquid Alloys [21]

Temperature (°C)	616-618				612-613				610-612			
Catalyst	K-In				K-Pb				Na-Pb			
Butene reacted	1-	trans-2-	cis-2-	1-	trans-2-	cis-2-	1-	trans-2-	cis-2-	1-	trans-2-	cis-2-
Total conv. (%)	30.3	28.0	31.7	27.3	26.4	27.5	24.3	33.1	39.0			
Butadiene yield (%)	10.3	8.7	10.3	9.4	8.8	9.6	10.2	7.8	9.0			
Selectivity (%)												
I[a]	34.0	31.1	32.4	34.4	33.4	34.9	42.0	23.7	23.0			
II[b]	39.4	58.4	59.4	34.8	54.3	55.8	30.0	68.1	72.4			
III[c]	26.6	10.5	8.2	30.8	12.3	9.3	28.0	8.2	4.6			
Overall selectivity for C$_4$ olefins (%)	73.4	89.5	91.8	69.2	87.7	90.7	72.0	91.8	95.4			

[a] The reaction I is defined by the dehydrogenation.
[b] The reaction II is defined by the isomerization.
[c] The reaction III is defined by the decomposition.

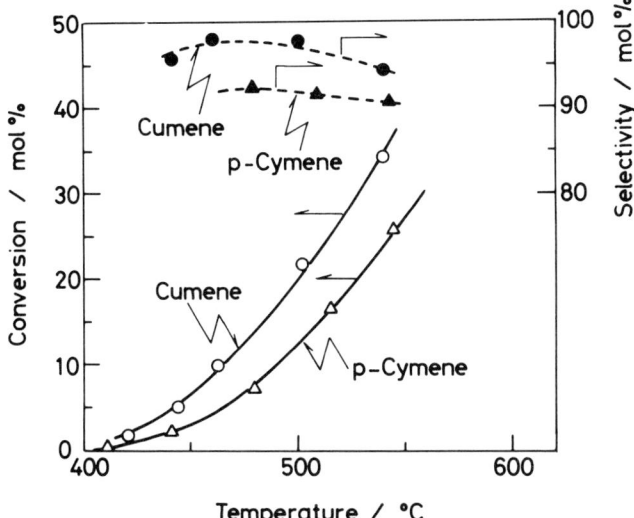

Fig. 7 The cumene dehydrogenating activity and the p-cymene dehydrogenating activity of the Na-Pb liquid alloy containing 50 atom% of Na [19].

To exemplify the catalytic activity, part of the experimental results are shown in Table 4 [21]. The yield of butadiene is appreciable, though yields of the isomerization products and the decomposition products are also appreciable.

Finally, it must be noted that the use of a diluent metal is indispensable, in order to obtain a stable and tractable catalyst. Although any one of the metals In, Tl, Sn, Pb, and Bi is available, In and Tl are most favorable because the resulting catalysts are very active and selective.

V. HYDROGEN TRANSFER REACTIONS

A. Reactions Between Alcohols and Ketones

A higher alcohol is either in a viscous state or in a solid state. Therefore, as mentioned in Section II of this chapter, we have to use an appropriate solvent to dissolve the reactant

TABLE 5 Hydrogen Transfer Reactions

Hydrogen donors	Hydrogen acceptors	Hydrogenateds			Refs.
HO~~~OH	>=O	>—OH			
	PhC(=O)CH₃	PhCH(OH)CH₃			
	cyclohexanone	cyclohexanol			[10]
	>=O (MEK)	>—OH			
⁻OH	>=O	>—OH			
⁻OH	cyclopentanone	cyclopentanol			
⁻OH	2-methylcyclohexanone	2-methylcyclohexanol	44.0[a]	56.0[b]	
3-hexanol (OH)	2-methylcyclohexanone	2-methylcyclohexanol	43.4	56.6	[13]
~~OH	3-methylcyclohexanone	3-methylcyclohexanol	58.2	41.8	
~~OH	4-methylcyclohexanone	4-methylcyclohexanol	44.5	55.5	

TABLE 5 (continued)

Hydrogen donors	Hydrogen acceptors	Hydrogenateds	Refs.
ethanol	hex-5-en-2-one	hex-5-en-2-ol	
butan-2-ol	4-methylcyclohex-2-enone	4-methylcyclohex-3-enol	
butan-2-ol	cyclohex-2-enone	cyclohex-2-enol (17.5 %[c]) / cyclohexanone (3.0 %[c])	[13]
pentan-2-ol	but-3-en-2-one	but-3-en-2-ol / butan-2-one	
benzyl alcohol	acetophenone / benzophenone / 4-methylbenzophenone	1-phenylethanol / diphenylmethanol / (4-methylphenyl)phenylmethanol	[1]

[a] Cis- selectivity.
[b] Trans- selectivity.
[c] Yield at 460°C.

alcohol to supply the reactant to the reaction apparatus. After some trials, acetone has been found to be most favorable, though hydrocarbon solvents such as cumene and p-cymene are also available in particular cases [1,16]. This situation has motivated studies on reactions between alcohols and ketones. Listed in Table 5 is the summary of the reactions studied. Among the reactions shown in this table, the reactions of terpene alcohols are worthy of a special comment in view of a practical importance.

Terpene alcohols and their derivatives are frequently utilized as raw materials for perfumes; citronellol and its dehydrogenated product, citronellal, are compounds of this sort. As we can see in Fig. 8 (dashed line), the liquid metal catalyst promotes the dehydrogenation of citronellol [1,11].

$$\text{citronellol-OH} \longrightarrow \text{citronellal} + H_2 \qquad (8)$$

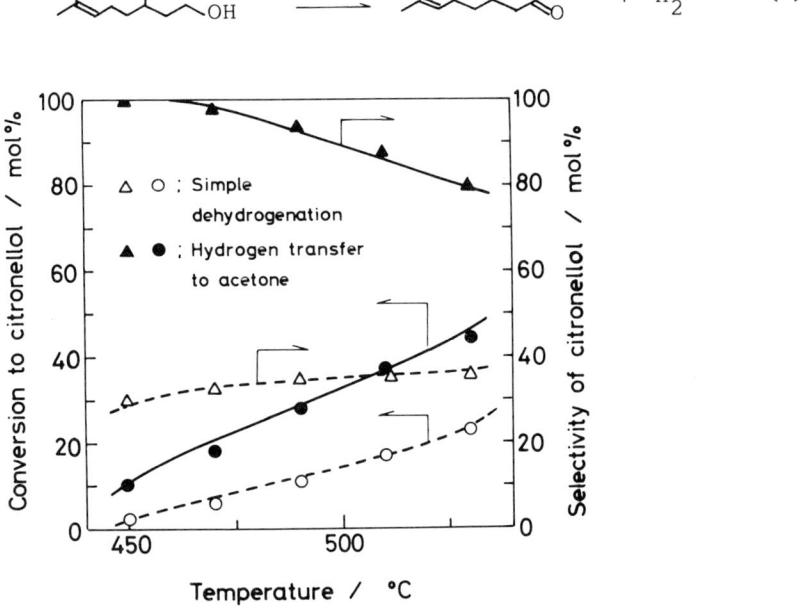

Fig. 8 A comparison of the apparent conversion and selectivity of the citronellol dehydrogenation with those of the hydrogen transfer from citronellol to acetone over Tl(liq.) [1,11].

However, the conversion and, in particular, the selectivity are not satisfactorily high. It is desirable to elevate the conversion with minimum losses of precious materials participating in the reaction. The experimental results shown in Fig. 8 (solid line) show the high conversion and the high selectivity of the following hydrogen transfer reaction [1,11].

(9)

$$\text{citronellol} + O=C(CH_3)_2 \longrightarrow \text{citronellal} + (CH_3)_2 CHOH$$

This demonstrates the utility of hydrogen transfer reactions for practical purposes. An efficient production of hydroxy citronellal from hydroxy citronellol is also possible if we use the hydrogen transfer reaction shown below [1,11].

(10)

$$\text{hydroxy citronellol} + (CH_3)_2 CO \longrightarrow \text{hydroxy citronellal} + (CH_3)_2 CHOH$$

A question may arise if the prominent effect of ketone is due entirely to the surface reaction. Miyamoto and Ogino [13], utilizing deuterium tracer techniques, have proved that the reaction between ethanol and cyclohexanone actually takes place over the In(liq.) catalyst surface.

Part of the reactions listed in Table 5 have been carried out to investigate stereochemical aspects of the catalysis. It has been inferred that the interaction between an alcohol molecule and a ketone molecule would be restricted by the flat surface of the liquid metal catalyst. Indeed the stereochemistry of the catalysis over the liquid metal has provided us with a very interesting problem. However, the discussion upon this this problem involves details of the reaction mechanism, and hence it is not pertinent to be described here. The stereochemistry of the catalysis will be discussed in the next chapter (Chap. 4, Section III).

TABLE 6 Hydrogen Transfer Reactions between 2-Butanol and Unsaturated Hydrocarbons

Acceptors	C=C (phenyl)	C=C (phenyl-OCH₃)	C=C (tolyl)	C-C=C (phenyl)	C=C-C (phenyl)	C-C=C (phenyl)
Conversion (mol %)	5	13	7	0.8	1.4	0.5
Temperature (°C)	510	530	540	518	510	546

Acceptors	C=C (cyclohexene)	indene			diphenylmethylene	diphenylethylene
Conversion (mol %)	1.3	1.1			4	2.2
Temperature (°C)	516	523			565	520

Source: Ref. 10.

Coal Liquefaction

Hydrogen transfer reactions involving unsaturated ketones as hydrogen acceptors are also studied [13]. As we can see in Table 5, the $>C = C<$ double bonds in β,γ- and γ,δ-unsaturated ketones are not hydrogenated, whereas those of the α,β-unsaturated ketone are hydrogenated.

B. Reactions between 2-Butanol and Unsaturated Aromatic Hydrocarbons

Generally the hydrogen transfer reactions between alcohols and unsaturated hydrocarbons over liquid metal catalysts are not marked. The conversion is low even when the reaction takes place [10]. Summarized in Table 6 are the reactions examined.

VI. COAL LIQUEFACTION

A. Liquefaction with Sn(liq.)

A coal liquefaction study with Sn(liq.) catalyst has been carried out by Weller et al in 1950 [31] and later by Ohuchi et al [32]. Systematic researches on this particular catalytic system were, however, initiated in 1981 [23]. The Sn(liq.) catalyst has been found to bring about various advantages compared with traditional solid-state catalysts: Information about the solid-state catalyst is abundant in literature, e.g., [33]. For instance, the Sn(liq.) survives severe liquefaction conditions without little loss of its original property. The catalyst metal can almost completely be recovered in the form shown in Fig. 9 [34]. This figure shows that the catalyst has undoubtedly been in the liquid state during the reaction: Tin shots have been charged into the reactor. The Auger electron spectroscopic result shown in Fig. 10 demonstrates that the catalyst has been attacked very little by other substances existing under the reaction conditions.

Part of the results of nonsolvent and Sn(liq.)-catalyzed liquefaction of coals of various ranks are shown in Fig. 11 [25]. It is clear that Sn(liq.) catalyzes the coal liquefaction. The experimental data for the Shin-Yubari coal are shown in Figs. 12

Fig. 9 An outward seeming of the tin catalyst after the use in the coal liquefaction for a 4-hr period at 400°C [34].

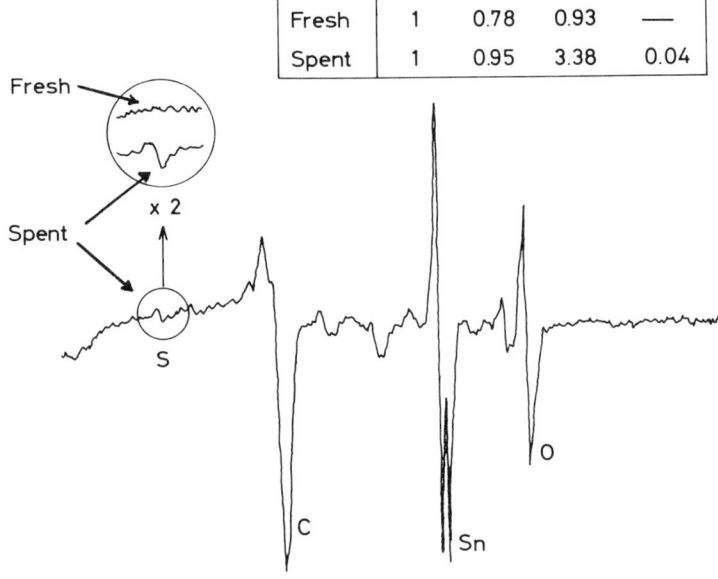

	Sn	O	C	S
Fresh	1	0.78	0.93	—
Spent	1	0.95	3.38	0.04

Fig. 10 Auger spectra of the Sn catalyst after use in coal liquefaction [34].

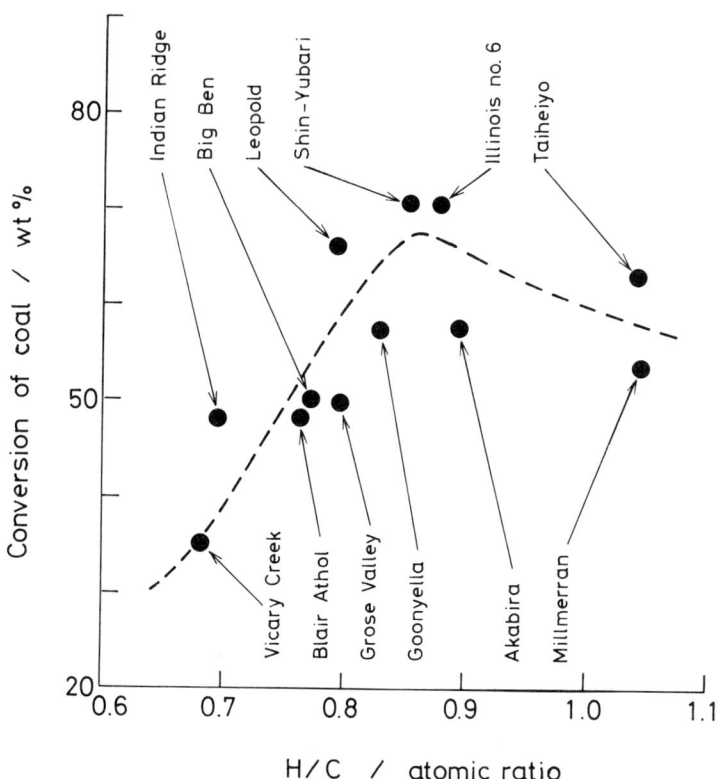

Fig. 11 Results of nonsolvent liquefaction reactions of coals of different ranks over the Sn(liq.) catalyst at 400°C for a 4-hr period using hydrogen (the initial pressure $p_{H_2}^0$ = 8 MPa). (H/C) Hydrogen to carbon atomic ratio adopted as the measure of the coal rank [25].

and 13 [23]. The former figure reveals that Sn(liq.) most effectively catalyzes the asphaltenes formation: Usually coal liquefaction products are successively fractionated into preasphaltenes, asphaltenes, oils-1, and oils-2 by Soxhlet extraction. The latter figure suggests that the polymerization of fragment radicals prevails at temperatures higher than 400°C.

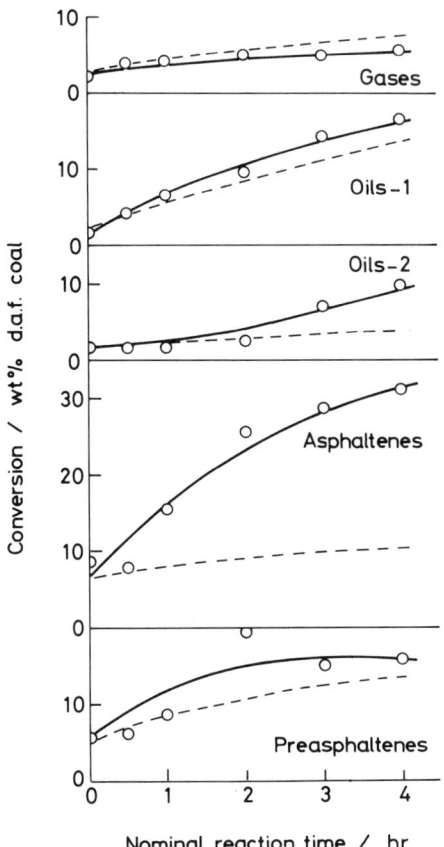

Fig. 12 Results of kinetic runs for the liquefaction of the Sin-Yubari coal over Sn(liq.) at 400°C using hydrogen ($p_{H_2}^0$ = 8 MPa): (——) with Sn(liq.) as a catalyst, (----) without catalyst [24].

The coal liquefaction is usually accompanied with product characterizations: elemental analysis, GPC, NMR, IR, FD-MS (field desorption mass spectrometry), and other techniques are used to obtain data necessary for determining structural parameters of the products [35]. Presented in Table 7 are the values

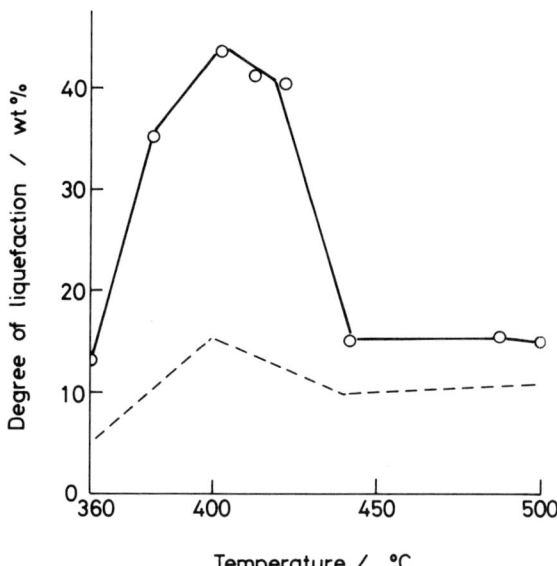

Fig. 13 The temperature dependence of the degree of liquefaction for the Shin-Yubari coal (period = 4 hr, $p^0_{H_2}$ = 8 MPs). (———) with catalyst, (----) without catalyst [23].

of the structural parameters of the products of Sn(liq.)-catalyzed liquefaction of the Shin-Yubari coal [36]. Although the purpose of evaluating the structural parameters is to obtain structural models of the products, it is not possible to make rigorous modelings with only the data given in the table. Therefore the modelings shown in Fig. 14 have to be regarded as very crude approximations [34]. Nevertheless, these modeling studies supplement experiments and enable us to speculate about the roles played by the Sn(liq.) catalyst. Figure 14 shows that the catalyst greatly promotes the intermediate asphaltenes formation while it promotes little the hydrogenation of asphaltenes.

An efficient coal liquefaction usually employs an appropriate organic solvent, e.g., tetralin as a good hydrogen-

TABLE 7 Structural Parameters of Liquefaction Products

Structural Parameters	Preasphaltenes	Asphaltenes	Oils-2	Oils-1
Molecular weight	900-1300	600-700	300	250, 340
Degree of aromatic ring condensation	4-6	4-6 (3-4)[a]	2	1
Degree of substitution	0.3-0.4	0.3-0.4 (0.3)	0.45	0.5
Average length of side-chain	3	2.5 (2.3-2.5)	2	2.5-3
Aromaticity	0.7-0.8	0.6-0.7 (0.7-0.8)	0.5-0.6	0.4-0.5
Total number of rings per molecule	15-19	10-12	3-4.5	2.9-3.8
Number of aromatic rings per molecule	12-15[b]	7.8[b]	1.3-2.5	1.4-2.0
Number of unit structures per molecule	2.7-5.0[b]	1.8-2.3[b]	1.4-1.7	1.6-1.9

[a]Values in parentheses are for the products obtained without catalyst.
[b]Peri-type structures are assumed.
Source: Ref. 36.

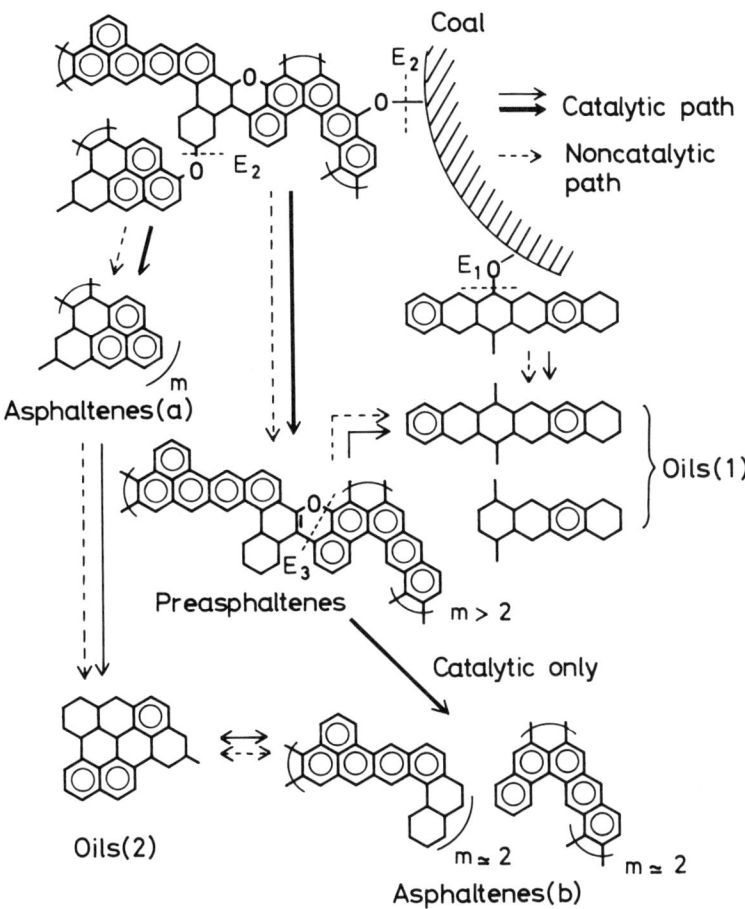

Fig. 14 A schematic reaction pathways for the liquefaction of the Shin-Yubari coal [34].

donating solvent [37]. It is expected that the solvent effect of tetralin would assist the catalytic effect of Sn(liq.). Indeed, as we can see in Fig. 15, the Sn(liq.) catalyst coupled with small amounts of tetralin brings about a higher yield of oils-1 and oils-2. A slight decrease in the asphaltenes formation in the later stage of the liquefaction (Fig. 15) appears

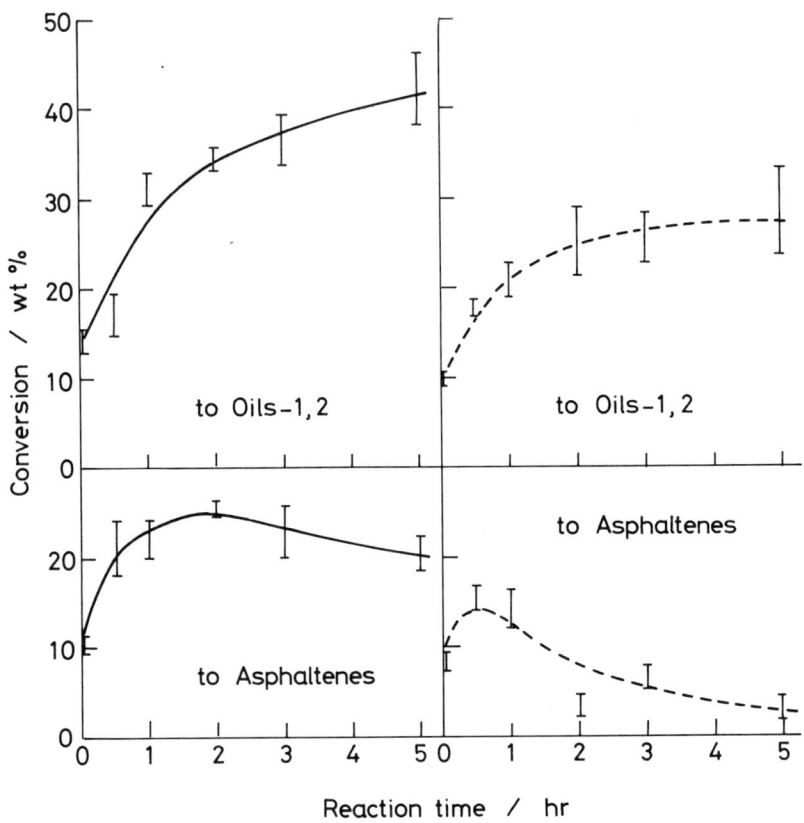

Fig. 15 Results of liquefaction of the Shin-Yubari coal with (———) and without (----) the Sn(liq.) catalyst in the presence of tetralin as a solvent (temperature = 420°C, $p^0_{H_2}$ = 4 MPa, solvent/coal = 1 wt ratio) [27].

to indicate that the hydrogenation of asphaltenes to form oils is really promoted. Cochran et al [37] have reported that Sn(liq.) catalyzes hydrogenation of fragment radicals of coal. There are other data that suggest that the surface of Sn(liq.) captures fragment radicals of coal and prevents them from polymerization [27]. However, it must be noted here that the contribution of the Sn(liq.) catalyst is observable only when the ratio

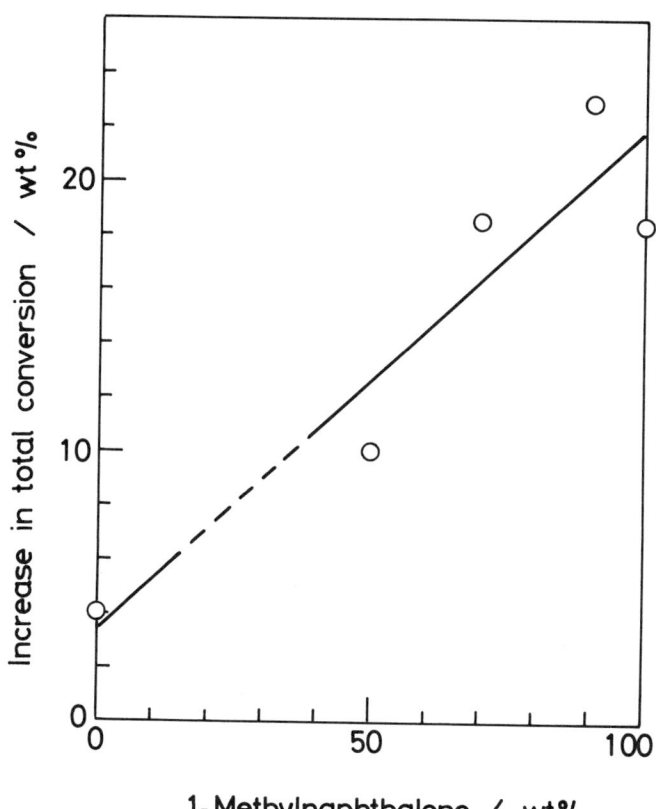

Fig. 16 The catalytic effect of Sn(liq.) upon the coal conversion as a function of the composition of the solvent composed of tetralin and naphthalene (temperature = 420°C, period = 1 hr, $p^0_{H_2}$ = 4 MPa) [26].

of tetralin/coal is less than 1 by weight. An excess in the tetralin addition smears the catalytic effect.

The Sn(liq.) catalyst exhibits another effect in the coal liquefaction in the 1-methylnaphthalene-tetralin mixed solvent. As we can see in Fig. 16 [26], the Sn(liq.) catalyst enhances the coal conversion as the concentration of 1-methylnaphthalene increases. It is known that 1-methylnaphthalene is a hydrogen-shuttling solvent [38]. Therefore, Fig. 29 shows that Sn(liq.) has promoted the H-shuttling effect of 1-methylnaphthalene. Since the H-shuttling involves radical reactions [38], the interaction between the Sn(liq.) catalyst and organic free radicals is suspected of causing the catalytic effect observed.

B. Liquefaction Activities of Other Liquid Metals

As we can see in Fig. 17, liquid metals other than Sn(liq.) exhibit coal liquefaction activities that correlate with the standard enthalpy of oxide formation (ΔH_f^0) of the catalyst metal. The volcano-shaped correlation strongly suggests that cleavages of any oxygen-containing linkages, in particular ether linkages [39,40], in the coal structure are playing important roles in the coal liquefaction. The reaction model shown in Fig. 18 [34] serves for the discussion on this problem, though it is much simplified.

The reaction model shown below (Fig. 18) gives the following expression for the observed rate constant k_{obs} [34,41],

$$k_{obs} = \frac{k_3}{(S-1)!} \left(\frac{E_0}{kT}\right)^{S-1} \exp(-E_0/kT) \qquad (11)$$

where E_0 is the activation energy, S is the effective internal freedom of the reacting molecule, k is Boltzmann's constant, T is the absolute temperature, and k_3 is the rate constant defined by Fig. 18. With Eq. (11) we can simulate the observed correlation between the coal conversion and ΔH_f^0 by assuming that $-\Delta H_f^0 = E_0$. The result of simulation is represented in Fig. 17 by a solid line. In this simulation, the internal degree of freedom S has been regarded as an adjustable parameter. The value of S suggests that many atoms (15-30 atoms) around the oxygen atom participate in the activation process.

C. Model Compound Reaction over Liquid Metal

The importance of studying radical reactions using appropriate model compounds is widely recognized in the field of coal

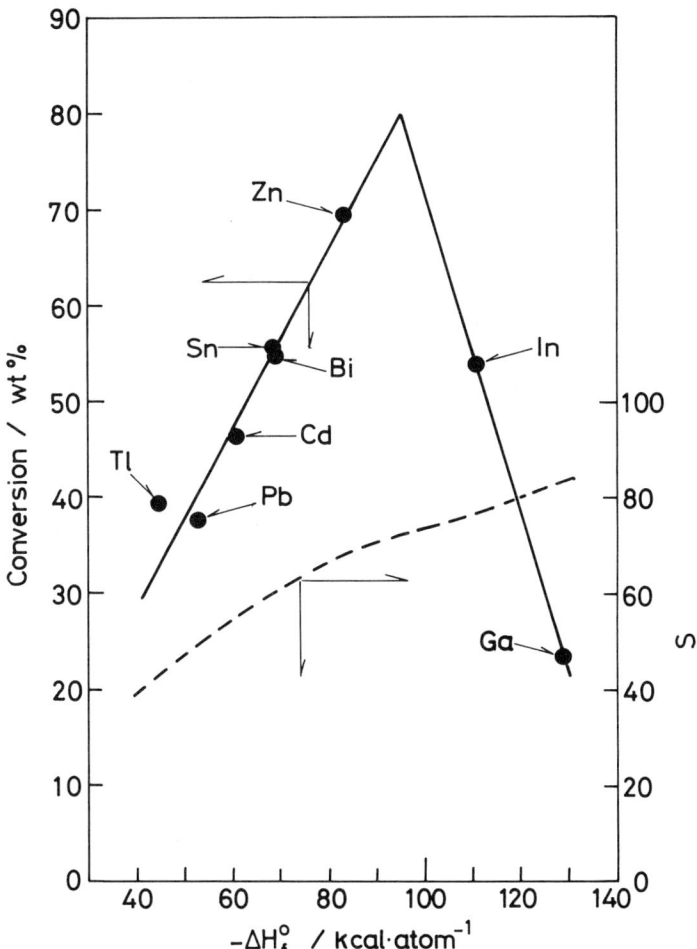

Fig. 17 Coal liquefaction activities of various liquid metals as a function of the standard enthalpy of oxide formation of the catalyst metal [22].

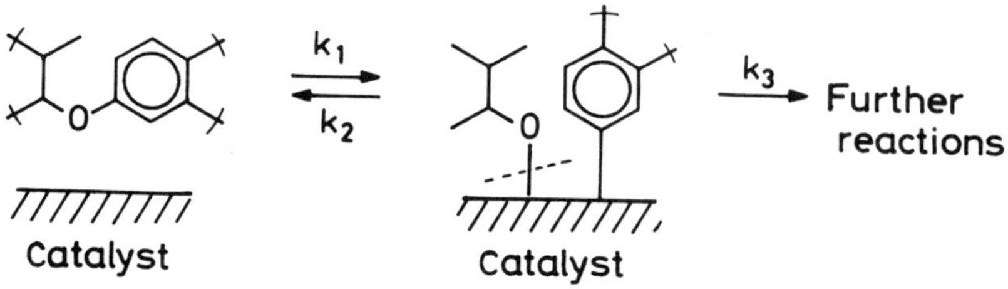

Fig. 18 A model for the interaction between the ether linkage in coal and the surface of liquid metal [34].

liquefaction chemistry [42-48]. Furthermore, as we have seen hitherto, important roles played by the interaction between liquid metals and free radicals have frequently been suggested. It appears interesting, therefore, to study any model compound reactions over the liquid metal.

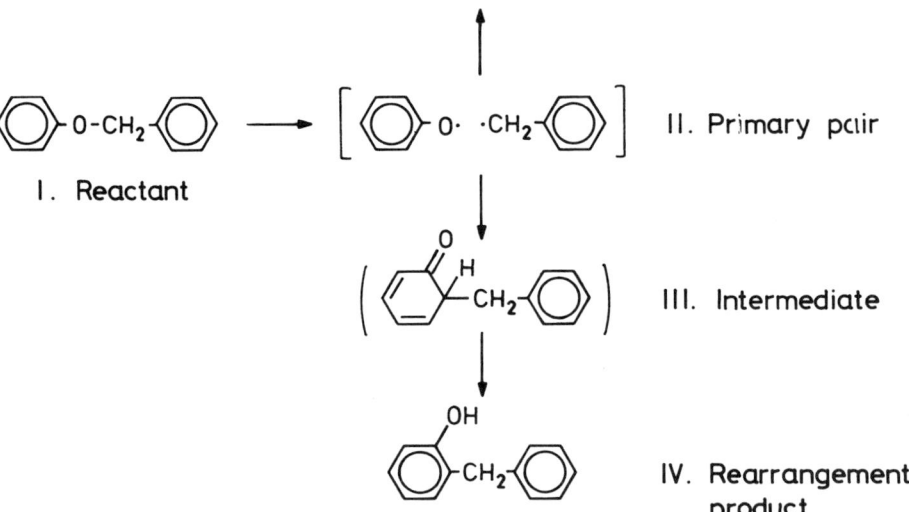

Fig. 19 A possible reaction scheme for the rearrangement of benzylphenylether over the Sn(liq.) catalyst [30].

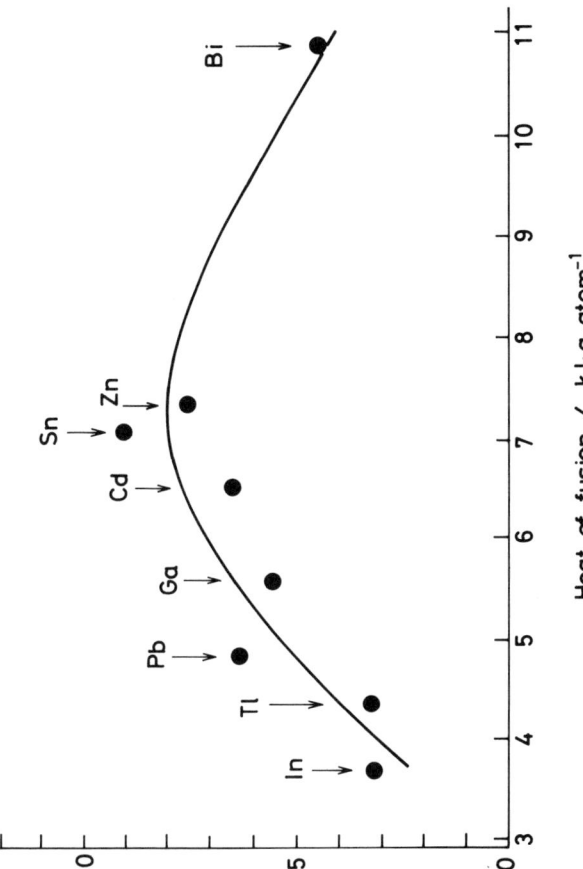

Fig. 20 A volcano-shaped relation between the heat of fusion of the catalyst metal and the catalytic activity for the isomerization of benzyl phenyl ether [51].

Recently, Ozawa et al [30] have reported the results of an experimental study on the decomposition of benzyl phenyl ether over the Sn(liq.) catalyst. According to this report, Sn(liq.) catalyzes only the isomerization pathway in the reaction scheme shown in Fig. 19. On the other hand, Sn(liq.) does little to promote the random mixture formation as well as phenol formation. The fact that the rearrangement is promoted by the Sn-(liq.) catalyst strongly suggests that the Sn(liq.) captures the primary pair on its surface by adsorption and provides the primary pair with changes to undergo isomerization. This resembles the cage effect observed for the reaction in solution [49,50].

The situation mentioned above does not change upon replacing Sn(liq.) by other liquid metals [51]. It has been reported that the rate of isomerization has a volcano-shaped correlation with the heat of fusion of the catalyst metal, as can be seen in Fig. 20. This suggests that a cluster of atoms of the catalyst surface participates in forming the transition state for the rearrangement.

D. Asphalt Digestion

Asphalt digestions are considered to have a common character with the coal liquefaction reaction: Both reactions involve heavy polymeric organic radicals in hydrogen-deficient states. Thus, it is expected that the liquid metal would show some catalytic activities in the asphalt digestion reaction, and research on the low-temperature digestion of Khafji crude have been carried out to clarify the catalytic effects of various liquid metals, i.e., Bi(liq.), Cd(liq.), Ga(liq.), In(liq.), Pb(liq.), and Sn(liq.) [28,29]. It is reported that the liquid metals effectively promote the molecular weight reduction with minimum gasification (less than 1 wt%). Although various data characterizing the reaction products have been reported, little mechanistic information is included.

VII. VARIOUS SURFACE REACTIONS REPORTED BY OTHER WORKERS

A. Research with Classical Techniques

Addison and his co-workers [52-75] have made intensive studies on the properties of alkali metals in the liquid state, and the results have already been summarized by Addison [76]. Except for the surface tension studies [52-54,57,58], their work is not always concerned with the problem of surface. Nevertheless, studies on reactions of various inorganic materials [69-71,76] and organic materials [64-66,68,75,76] deserve special attention.

In the discussion of the initial stage of a reaction between the liquid metal and a particular reactant, adsorption such as

$$Li(liq.) + N_2 \rightarrow \underset{Li \quad Li}{N-N} \rightarrow Li_3N \qquad (12)$$

or

$$\underset{-Na-}{\overset{HCl}{|}} \rightarrow \underset{-Na-}{\overset{HCl}{|}}\overset{e}{\searrow} \rightarrow \underset{-Na^+}{\overset{Cl^-}{|}} \quad \overset{H}{|} \qquad (13)$$

is frequently taken into account [64,76]. For instance, the reaction between Na(liq.) and propyne [70] is reported to involve the following adsorption and an electron transfer step

$$\underset{-Na-}{\overset{MeC\equiv CH}{|}} \rightarrow \underset{-Na-}{\overset{MeC\equiv CH}{|}}\overset{e}{\searrow} \rightarrow \underset{-Na^+}{\overset{[MeC\equiv C]^-}{|}} \quad \overset{H}{|} \quad \overset{MeC\equiv CH}{\diagup} \qquad (14)$$

$$\downarrow$$

$$\underset{-Na^+}{\overset{[MeC\equiv C]^-}{|}} + MeCH=CH_2$$

Acetylene reacts similarly [70], though two successive electron transfers are considered in this case;

$$\begin{array}{c} HC\equiv CH \\ \vdots \\ -Na- \end{array} \longrightarrow \begin{array}{c} HC\equiv CH \\ \vdots \\ -Na- \end{array} \xrightarrow{e} \longrightarrow \begin{array}{c} [HC\equiv C]^- \quad H \\ | \quad | \\ -Na^+ \end{array} \xrightarrow{e} \longrightarrow \begin{array}{c} [C\equiv C]^{2-} \quad H \quad H \\ | \quad | \quad | \\ -Na^+\!-Na^+ \end{array} \qquad (15)$$

Part of hydrogen atoms adsorbed react with surface acetylene to form ethylene and other atoms dissolve into the bulk of Na(liq.).

Although ethylene can not be adsorbed by Na(liq.) from a mixture of acetylene and ethylene, the adsorption becomes possible when small amount of Ba is added to Na(liq.) [70]

$$C_2H_4 \longrightarrow C_2H_4 \xrightarrow{e} [H_2C=CH]^- \xrightarrow{H} \xrightarrow{e} [HC=CH]^{2-} \quad H \quad H \downarrow$$

$$\xleftarrow{[HC\equiv C]^- \quad H}_{e} \xleftarrow{HC\equiv CH}_{e} \xleftarrow{[HC\equiv CH]}_{2e} \xleftarrow{H_2}$$

$$\downarrow \quad H \quad H \quad [C\equiv C]^{2-} \qquad (16)$$

The existence of surface ethylene and surface hydrogen atoms results in a ethane formation.

Pure ethylene can be adsorbed by Na(liq.) and by K(liq.) as well [76] in the manner similar to that mentioned above. The reactivity of K(liq.) is superior to that of Na(liq.). The reaction of ethylene and Na-Cs(6.3 atom%) liquid alloy is interesting in that significant amounts of n-butane are formed at lower temperatures [76]. This suggests that strongly adsorbed or long-lived $C_2H_5\cdot$ radicals exist on the alloy surface. Li(liq.) absorbs almost whole hydrogen formed from the interaction between the surface and adsorbed ethylene [76].

Various Surface Reactions

A strong electron donating power of a molten alkali metal to multiple bonds of organic compounds has also attracted a special attention of Friedman et al [77]. He assumed that the following type of catalysis would occur over the alkali metal M:

$$\text{CH}_2=\text{CH}_2 + 2M \longrightarrow \text{M-CH}_2-\text{CH}_2-\text{M} \xrightarrow{+2H_2} \text{CH}_3-\text{CH}_3 + (2MH \rightleftharpoons 2M + H_2) \quad (17)$$

$$\xrightarrow{+H_2} \text{CH}_3-\text{CH}_3 + 2M \quad (18)$$

and

$$MH + \text{CH}_2=\text{CH-CH}_3 \longrightarrow \text{M-CH}_2-\text{CH}_2-\text{CH}_3 \xrightarrow{+H_2} \text{CH}_3-\text{CH}_2-\text{CH}_3 + MH \quad (19)$$

On the basis of this assumption, he has examined a high-pressure hydrogenation of polynuclear aromatic hydrocarbons over the alkali metal catalyst as well as over the binary alloy catalyst containing alkali metals as components.

The experimental results have shown that both Na and K are good catalysts for the hydrogenation of polynuclear aromatic hydrocarbons, as expected by Friedman. However, the activity of Li has been found to be smaller than that of Na, suggesting a lower dispersion of Li in the reaction mixture.

Binary alloys such as Na-K, Na-Rb, and Na-Cs have also exhibited high catalytic activities: Most of these catalysts have been prepared in situ from Na and an appropriate inorganic compound such as KOH, K_2CO_3, or Rb_2CO_3. The order of the catalytic activity is Na-Rb > Na-Cs > Na-K.

The hydrogenation reactions mentioned above have been carried out well above the melting points of the catalysts. However it is not clear whether the catalyst has been in the liquid state under the reaction conditions. In the case of naphthalene hydrogenation, the catalyst (Na-K) has been recovered as a white solid suspended in the solvent phase (toluene) of the reaction products.

B. Research with Special Techniques

An interesting aerodynamic levitation reactor, as illustrated in Fig. 21, has been reported by Winborne [78]. In this reactor a spherical drop of a liquid metal is levitated by a gas jet from a specially designed nozzle and can react with an active component contained in the levitation gas (Ar). Although this reactor has been made to study the reaction of Al(liq.) with F_2 gas, the results have not yet been published.

Two recent reports of Balooch et al [79,80] are also worthy of special mention. Balooch and his co-workers have measured the rate of reaction between the liquid metal and Cl_2 using a modulated molecular beam technique coupled with an UHV (ultrahigh vacuum) technique (the base pressure in the target chamber is 5×10^{-9} Torr). The main part of the molecular beam chamber is outlined in Fig. 22. In addition to this apparatus, a separate

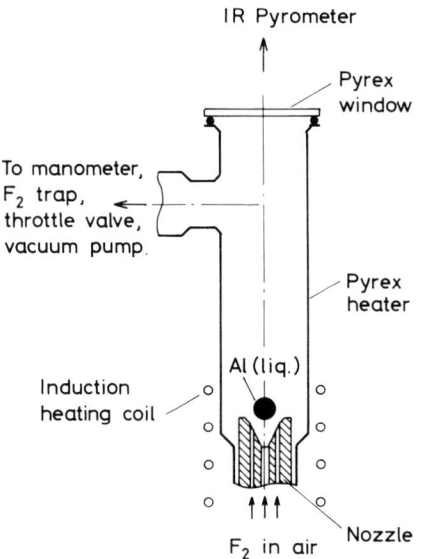

Fig. 21 An aerodynamic levitation technique for studying the reaction between Al(liq.) and F_2 gas [77].

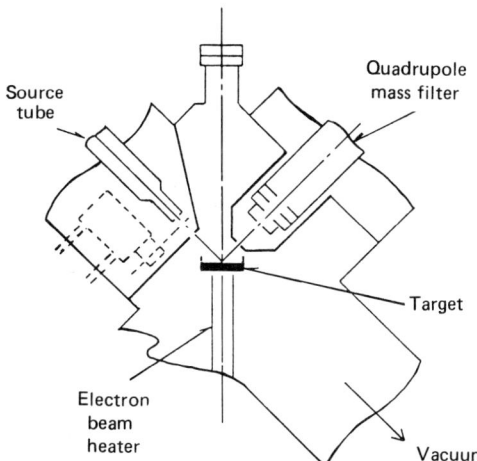

Fig. 22 The central part of a molecular beam apparatus for studying the interaction between the liquid metal and Cl_2 [78].

auxiliary reaction apparatus equipped with an AES (auger electron spectroscopy) surface analyzer has been used to monitor the sample surface during the reaction. Indium (In) and lead (Pb) have been chosen as samples because these metals melt at low temperatures and have low vapor pressures (10^{-8} Torr at 770 K for In and 617 K for Pb). The important information obtained in these studies are given below.

1. The sample surface can be regarded as a monolayer of chloride when the temperature is well below the melting point. On increasing temperature, the concentration of the surface chloride gradually decreases, and, on approaching the melting point, it decreases rapidly. The decreasing tendency of the surface concentration of the chloride continues after the melting, though it turns to level off at high temperatures.

2. The mode of disappearance of the surface chloride of In at the melting point differs from that of Pb. In the case of In, the surface InCl layer is broken on melting and desorbs

from the surface as InCl vapor, leaving island-like fragments of chloride on the liquid surface. On the other hand, the PbCl layer disappears from the surface of Pb partly by diffusion into the bulk of the Pb(liq.) and partly by the reaction with Cl to form $PbCl_2$, which desorbs easily.

3. Either for In or for Pb the rate of reaction with Cl_2 molecular beam is greatly enhanced on melting: The reactivity change is, nevertheless, continuous.

4. The reaction between In(liq.) and Cl_2 can be expressed as follows

$$Cl_2(g) \longrightarrow 2Cl_{ads} \tag{20}$$

$$Cl_{ads} + In \longrightarrow InCl \tag{21}$$

$$InCl \longrightarrow InCl(g) \tag{22}$$

The formation of $InCl_2$ takes place only on the surface covered by InCl

$$Cl_{ads} \text{ (on InCl)} + InCl \longrightarrow InCl_2(g) \tag{23}$$

On the other hand, only $PbCl_2$ is formed by the reaction of Pb(liq.) with Cl_2

$$Cl_{ads} + PbCl_{ads} \longrightarrow PbCl_2(g) \tag{24}$$

and

$$PbCl_{ads} + Cl_2(g) \longrightarrow PbCl_2(g) + Cl_{ads} \tag{25}$$

Finally, there remains one interesting paper reported by Nilsson and Rabinovich [81]. This work is an extension of the accommodation coefficient studies [82-84] for gas-solid systems to the Sn(liq.)-cyclohexene system. The probability of excitation of cyclohexene molecules to energy levels above the threshold of conversion to 1,3-butadiene in one collision with the surface has been measured. It is reported that the accommodation coefficient is continuous at the melting point and that the Sn(liq.)

surface is much like the solid surface with respect to the energy transfer. It is inferred that island-like Sn(II) oxide fragments would be floating on the Sn(liq.) surface.

REFERENCES

1. Y. Saito, Ph.D. thesis, Tohoku, University, 1978.
2. Y. Saito, A. Miyamoto, and Y. Ogino, *Kogyo Kagaku Zassi*, 74, 1521 (1971).
3. K. Kashiwadate, Y. Saito, A. Miyamoto, and Y. Ogino, *Bull. Chem. Soc. Jpn.*, 44, 3004 (1971).
4. Y. Saito, N. Hiramatsu, N. Kawanami, and Y. Ogino, *Bull. Jpn. Petrol. Inst.*, 14, 169 (1972).
5. A. Miyamoto and Y. Ogino, *J. Catal.*, 27, 311 (1972).
6. Y. Saito, F. Miyashita, and Y. Ogino, *J. Catal.*, 36, 67 (1975).
7. Y. Ogino, Y. Saito, and K. Okano, *Semi-annu. Rep. Asahi Glass Found. Contrib. Ind. Technol.*, 22, 37 (1972).
8. K. Okano, Y. Saito, and Y. Ogino, *Bull. Chem. Soc. Jpn.*, 45, 69 (1972).
9. G. M. Schwab and H. H. Martin, *Z. Elektrochem.*, 43, 610 (1937).
10. K. Takahashi, Ph.D. thesis, Tohoku University, 1979.
11. Y. Saito and Y. Ogino, *Nippon Kagaku Kaishi*, 1018 (1976).
12. Y. Saito, H. Yoshida, A. Miyamoto, T. Yokoyama, and Y. Ogino, *J. Catal.*, 55, 36 (1978).
13. A. Miyamoto and Y. Ogino, *J. Catal.*, 43, 143 (1976).
14. H. Sugawara, Ms. thesis, Tohoku University, 1980.
15. Y. Saito and Y. Ogino, *J. Catal.*, 55, 198 (1978).
16. H. Sugawara and Y. Ogino, *J. Chem. Soc., Faraday Trans.*, 1, 78, 1079 (1982).
17. K. Takahashi and Y. Ogino, *Chem. Lett.*, 423 (1978).
18. K. Takahashi and Y. Ogino, *Chem. Lett.*, 549 (1978).
19. K. Honda, K. Takahashi, Y. Saito, and Y. Ogino, *Chem. Lett.*, 693 (1978).
20. K. Honda and Y. Ogino, *J. Chem. Soc. Chem. Comm.*, 332 (1980).

21. K. Honda, K. Takahashi, and Y. Ogino, *Chem. Lett.*, 553 (1980).
22. M. Matsuura, S. Matsunaga, S. Ozawa, and Y. Ogino, *J. Chem. Soc. Chem. Comm.*, 721 (1981).
23. M. Matsuura, S. Matsunaga, S. Ozawa, and Y. Ogino, *Fuel*, 62, 407 (1983).
24. S. Ozawa, M. Matsuura, S. Matsunaga, and Y. Ogino, *Fuel*, 63, 719 (1984).
25. Y. Ogino, S. Ozawa, and M. Komiyama, *Res. Coal Liquef. Gasif., Rep. Special Project Research on Energy Under Grant in Aid of Scientific Research of the Ministry of Education and Culture*, SPEY 12, 1984, p. 101.
26. S. Ozawa, H. Yamazaki, and Y. Ogino, *Sekiyu Gakkaishi*, 28, 349 (1985).
27. Y. Ogino, S. Ozawa, and K. Ishikawa, in *Fuel Processing Technology* (Y. Sanada, ed.), Elsevier Scientific Publishing Company, Amsterdam, 1985.
28. M. Komiyama, S. Nojima, and Y. Ogino, *Fuel*, 61, 542 (1982).
29. M. Komiyama, S. Nojima, and Y. Ogino, in *Characterization of Heavy Crude Oils and Petroleum Residues* (Proc. Intern. Symp., Lyon, France), Editions Technip, Paris, 1984, p. 428.
30. S. Ozawa, T. Suenaga, and Y. Ogino, *Fuel*, 64, 712 (1985).
31. S. Weller, M. G. Pelipetz, S. Friedman, and H. H. Storch, *Ind. Eng. Chem.*, 42, 330, 334 (1950).
32. K. Makino, N. Orihara, and K. Ohuchi, *Nenryo Kyokaishi*, 57, 718 (1978).
33. J. A. Cusumano, R. A. Dalla Betta, and R. B. Levy, in *Catalysis in Coal Conversion*, Academic Press, New York, San Francisco, London, 1978.
34. Y. Ogino, S. Ozawa, M. Komiyama, and S. Matsunaga, in *China-Japan-USA Symposium on Heterogeneous Catalysis Related to Energy Problem* (Aug. 31-Sept. 2, Dalian, China), 1982, A17J.
35. J. K. Brown, W. R. Lander, and N. Sheppard, *Fuel*, 39, 79, 87 (1960).
36. S. Matsunaga, Ms. thesis, Tohoku University, 1983.
37. S. J. Cochran, M. Hatswell, W. R. Jackson, and F. P. Larkins, *Fuel*, 61, 831 (1982).
38. D. D. Whitehurst, M. Farcasiu, and T. O. Michell, in *Coal Liquefaction*, Academic Press, New York, London, Toronto, Sydney, San Francisco, 1980, p. 274.

References

39. Y. Kamiya, *Nenryo Kyokaishi*, *57*, 12 (1978).
40. H. Ohta and Y. Kamiya, *J. Syn. Org. Chem. Jpn.*, *38*, 912 (1980).
41. P. J. Robinson and K. A. Holbrook, in *Unimolecular Reaction*, Wiley Interscience, 1972, pp. 15-27.
42. L. W. Vernon, *Fuel*, *59*, 102 (1980).
43. M. L. Poutsma, *Fuel*, *59*, 335 (1980).
44. R. H. Schlosberg, T. R. Ashe, R. J. Pancirov, and M. Donaldson, *Fuel*, *60*, 155 (1981).
45. J. M. L. Penninger and K. Versluis, *Fuel*, *61*, 283 (1982).
46. M. L. Poutsma and C. W. Dyer, *J. Org. Chem.*, *18*, 3367 (1982).
47. M. Siskin and T. Aczel, *Fuel*, *62*, 1321 (1983).
48. L. R. Rudnick and D. Tueting, *Fuel*, *63*, 153 (1984).
49. D. Ellus, *Adv. Photochem.*, *8*, 109 (1970).
50. N. Shimamura and A. Sugimori, *Bull. Chem. Soc. Jpn.*, *44*, 281 (1971).
51. S. Ozawa, K. Sasaki, and Y. Ogino, in preparation.
52. C. C. Addison, D. H. Kerridge, and J. Lewis, *J. Chem. Soc.*, 2861 (1954).
53. C. C. Addison, W. E. Addison, D. H. Kerridge, and J. Lewis, *J. Chem. Soc.*, 2262 (1955).
54. C. C. Addison, W. E. Addison, and D. H. Kerridge, *J. Chem. Soc.*, 3047 (1955).
55. C. C. Addison, W. E. Addison, D. H. Kerridge, and J. Lewis, *J. Chem. Soc.*, 1454 (1956).
56. C. C. Addison, E. Iberson, and J. A. Manning, *J. Chem. Soc.*, 2699 (1962).
57. C. C. Addison, J. M. Coldrey, and W. D. Halstead, *J. Chem. Soc.*, 3868 (1962).
58. C. C. Addison, J. M. Coldrey, and R. J. Pulham, *J. Chem. Soc.*, 1227 (1963).
59. C. C. Addison and R. J. Pulham, *J. Chem. Soc.*, 1232 (1963).
60. C. C. Addison, R. J. Pulham, and R. J. Roy, *J. Chem. Soc.*, 4895 (1964).
61. C. C. Addison, R. J. Pulham, and R. J. Roy, *J. Chem. Soc.*, 116 (1965).
62. C. C. Addison, M. G. Barker, and R. J. Pulham, *J. Chem. Soc.*, 4483 (1965).

63. C. C. Addison, G. K. Creffield, P. H. Hubberstey, and R. J. Pulham, *J. Chem. Soc. (A)*, 1482 (1969).

64. C. C. Addison and B. M. Davies, *J. Chem. Soc. (A)*, 1822 (1969).

65. C. C. Addison and B. M. Davies, *J. Chem. Soc. (A)*, 1828 (1969).

66. C. C. Addison and B. M. Davies, *J. Chem. Soc. (A)*, 1831 (1969).

67. C. C. Addison, G. K. Creffield, P. H. Hubberstey, and R. J. Pulham, *J. Chem. Soc. (A)*, 1393 (1971).

68. R. J. Pulham, *J. Chem. Soc. (A)*, 1389 (1971).

69. C. C. Addison, M. R. Hobdel, and R. J. Pulham, *J. Chem. Soc. (A)*, 1700 (1971).

70. C. C. Addison, M. R. Hobdel, and R. J. Pulham, *J. Chem. Soc. (A)*, 1704 (1971).

71. C. C. Addison, M. R. Hobdel, and R. J. Pulham, *J. Chem. Soc. (A)*, 1708 (1971).

72. P. Hubberstey and R. J. Pulham, *J. Chem. Soc. Dalton*, 819 (1972).

73. P. Hubberstey and R. J. Pulham, *J. Chem. Soc. Dalton*, 1541 (1974).

74. M. R. Hobdel and A. C. Whittingham, *J. Chem. Soc. Dalton*, 1591 (1975).

75. G. Parry and R. J. Pulham, *J. Chem. Soc. Dalton*, 1915 (1975).

76. C. C. Addison, in *The Chemistry of the Liquid Alkali Metals*, John Wiley and Sons, Chichester, New York, Brisbane, Toronto, Singapore, 1984.

77. S. Friedman, M. L. Kaufman, and I. Wender, *J. Org. Chem.*, 36, 694 (1971).

78. D. A. Winborne, P. C. Nordine, D. E. Rosner, and N. F. Marley, *Metall. Trans.*, 7B, 711 (1976).

79. M. Balooch, W. J. Siekhaus, and D. R. Olander, *J. Phys. Chem.*, 88, 3521 (1984).

80. M. Balooch, W. J. Siekhaus, and D. R. Olander, *J. Phys. Chem.*, 88, 3526 (1984).

81. W. B. Nilsson and B. S. Rabinovich, *Langmuir*, 1, 71 (1985).

82. D. F. Kelley, B. D. Barton, L. Zalotai, and B. S. Rabinovich, *J. Chem. Phys.*, 71, 538 (1979).

83. D. F. Kelley, L. Zalotai, and B. S. Rabinovich, *J. Chem. Physics*, *46*, 379 (1980).

84. W. Yuan and B. S. Rabinovich, *J. Phys. Chem.*, *87*, 2167 (1983).

4
Advanced Problems in Liquid Metal Catalysis

I.	Introduction	74
II.	Kinetics	75
	A. Dehydrogenation of Alcohols	75
	B. Kinetics for Other Reactions	79
III.	Stereochemistry	87
IV.	Problems in Catalysis by Liquid Alloy	94
	A. Catalytic Activity and Alloy Composition	94
	B. Atomic Distribution in Liquid Alloy	96
	C. Surface Composition	101
V.	Electronic Aspect of Catalysis	111
	A. Qualitative Results	111
	B. Semi-Quantitative Results	115
	References	122

I. INTRODUCTION

We now face the problem of disclosing the reaction mechanism and the catalyst structure, and we hope to draw a clear picture of liquid metal catalysis at the atomic and electronic levels. Of course, the present situation regarding fundamental research on liquid metal catalysis is far from the ultimate goal, and many

Introduction

problems are still open to further study. Nevertheless, considerable progress has been made in many respects, and the purpose of this chapter is to present a review of these results.

II. KINETICS

A. Dehydrogenation of Alcohols

Kinetic studies of the liquid metal catalysis were initiated with an alcohol-liquid metal reaction system [1-4]. The technique necessary for studying the kinetics have already been stated in Chapter 2.

Alcohol dehydrogenation obeys two different kinetics, depending on the reaction pressure. Under pressures higher than ~0.1 atm, the reaction rate is expressed by

$$r = \frac{kK_A p_A}{(1 + K_A p_A)} \tag{1}$$

where r is the reaction rate, k is the rate constant, K_A is the adsorption equilibrium constant for the reactant alcohol, and p_A is a pressure of the reactant vapor.

On the other hand, when the pressure is lower than ~0.1 atm, the reaction rate is expressed by

$$r = \frac{k(K_A p_A)^2}{(1 + K_A p_A)^2} \tag{2}$$

The following reaction scheme has been considered [3,4] to explain the experimental results mentioned above;

$$A \xrightarrow{K_A} A^* \quad \text{(Adsorption)} \tag{3}$$

$$A^* + A^* \xrightarrow{k_1} A^{**} + A^* \quad \text{(Activation by a surface bimolecular collision)} \tag{4}$$

$$A^{**} + A^* \xrightarrow{k_2} A^* + A^* \quad \text{(Deactivation by a surface bimolecular collision)} \tag{5}$$

$$A^* \xrightarrow{k_3} \text{Products} \quad \text{(Surface unimolecular decomposition of the activated admolecule)} \quad (6)$$

Under the assumption that the Langmuir type of adsorption isotherm is applicable to the adsorption of A (the step expressed by Eq. 3) and that the stationary-state treatment is applicable, we obtain the following equation;

$$r = \frac{k_1 [K_A P_A / (1 + K_A P_A)]^2}{[1 + \{(k_2/k_3) K_A P_A / (1 + K_A P_A)\}]} \quad (7)$$

When the pressure is low, we can assume that the surface is sparsely covered by the reactant molecules, i.e.,

$$\frac{(k_2/k_3) K_A P_A}{(1 + K_A P_A)} \ll 1 \quad (8)$$

By combining Eq. 7 with Eq. 8 we obtain

$$r = \frac{k_1 (K_A P_A)^2}{(1 + K_A P_A)^2} \quad (9)$$

which agrees with the experimental rate equation (Eq. 2). On the other hand, when the pressure is high, we can assume that the surface is almost saturated with the adsorbed molecules and hence

$$\frac{(k_2/k_3) K_A P_A}{(1 + K_A P_A)} \gg 1 \quad (10)$$

By combining Eq. 7 with Eq. 10 we obtain

$$r = \frac{k_1 (k_3/k_2)(K_A P_A)}{1 + K_A P_A} \quad (11)$$

which agrees with the experimental rate equation (Eq. 1).

The reaction scheme mentioned above assumes that activations of molecules are caused by the surface bimolecular collisions. This is only possible when the admolecules are mobile. Indeed, as we can see in Table 1, the experimentally determined value of

TABLE 1 Values for the Theoretical and Experimental Adsorption Equilibrium Constant K_A^0 [a]

Co-area, σ (Å^2)	Vibrational frequency, $\tilde{\nu}$ (cm^{-1})	$K_{A,calc}^0$ [b] (atm^{-1})	$K_{A,obs}^0$ (atm^{-1})
25	1	1.08×10^{-2}	
	10	1.06×10^{-3}	
	70	1.64×10^{-4}	
50	1	2.18×10^{-2}	
	10	2.12×10^{-3}	6.0×10^{-2}
	70	3.28×10^{-4}	
100	1	4.32×10^{-2}	
	10	4.24×10^{-3}	
	70	6.56×10^{-4}	

[a] K_A^0 is defined by $K_A = K_A^0 \exp(q/RT)$, where q is the heat of adsorption.
[b] Mean values (range 400–600°C).
Source: Refs. 3 and 4.

the adsorption equilibrium constant agrees with the theoretical value calculated by assuming a mobile adsorption and by applying the statistical theory of adsorption [5], i.e.,

$$K_A = \frac{hf_{vib} \sigma \exp(q/RT)}{(kT)(2\pi mkT)^{1/2}} \qquad (12)$$

where h is Planck's constant, k is Boltzmann's constant, m is the molecular mass of adsorbate, q is the heat of adsorption, R is the gas constant, T is the absolute temperature, f_{vib} is the partition function for a vibrational motion perpendicular to the catalyst surface, and σ is the co-area of the admolecule. The co-area σ and the vibrational frequency $\tilde{\nu}$ have been estimated taking the values reported by Kemball [6] into consideration.

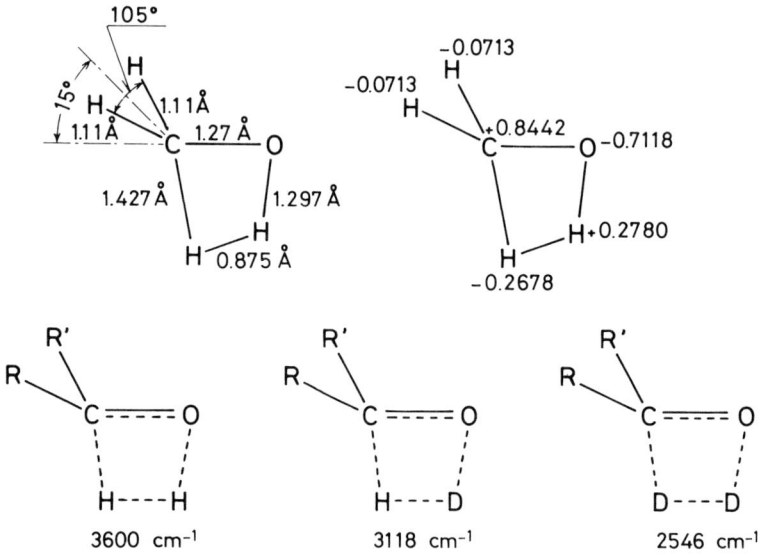

Fig. 1 Various aspects of the transition states in the dehydrogenation of alcohols over In(liq.) [2,8].

It has also been found that the heat of adsorption q is a reasonable order of magnitude as a mobile adsorption compared with the values reported by Kemball [6]: The experimental values are 9 kcal/mol for 2-propanol [3] and 12 kcal/mol for 2-butanol. The validity of the mobile adsorption assumption and that of the resulting surface bimolecular activation mechanism have also been proved by analyzing the rate constant [1-4].

Another aspect of the kinetics and mechanism of the liquid metal catalysis can be revealed by the analysis of the kinetic isotope effects [2,7,8]. The following data serve for the analysis

X_{d-0}/X_{d-1} = 1.74   methanol-In(liq.)

X_{d-1}/X_{d-4} = 1.37   methanol-In(liq.)   (13)

X_{d-1}/X_{d-6} = 1.22   ethanol-In(liq.)

$X_{d-0}/X_{d-1} = 1.66$ ethanol-In(liq.), and
2-propanol-In(liq.) (13) cont.
$X_{d-1}/X_{d-8} = 1.22$ 2-propanol-In(liq.)

where X denotes the conversion of the reactant alcohol and the subscript d-n denotes that the deuteroalcohol contains n deuterium atoms.

The theoretical meaning of the kinetic isotope effects are well summarized in the literature [9,10] and the details of the analysis of the data shown above are reported in the original papers [2,8]. Thus, only the transition-state models derived from the analysis of the isotope effects are shown in Fig. 1. It must be pointed out that the transition state models are proper only when the potential energy-surface for the gas-phase reaction can well simulate that of the surface reaction [8].

B. Kinetics for Other Reactions

1. *Dehydrogenation of Tetralin*

The dehydrogenation of tetralin is well catalyzed by Te(liq.), as we have seen in Chapt. 3, IV. However, the kinetics of this reaction is somewhat complicated owing to the following reasons [11,12]:

1. The formation of dihydro-naphthalene is significant, suggesting that the reaction is consecutive, i.e.,

$$\text{tetralin} \xrightarrow{k_1} \text{dihydronaphthalene} + H_2 \xrightarrow{k_2} \text{naphthalene} + 2H_2 \quad (14)$$

2. The tetralin dehydrogenation is catalyzed not only by Te(liq.) but also by Te(vapor), and, in addition, pyrolytic dehydrogenation takes place.

Considering the situation mentioned above, Takahashi and Ogino [11,12] have applied the experimental techniques described in Chapt. 2, II and have found that the tetralin dehydrogenation

TABLE 2 Comparison[a] between the Intrinsic Catalytic Activity[b] of Te(liq.) and That of Te(vapor)

Temperature (°C)	Ratios of rate constants[c]	
	$k_{1,liq}/k_{1,vap}$	$k_{2,liq}/k_{2,vap}$
570	24.1	27.8
560	18.7	21.7
545	17.9	16.8
530	11.7	9.1
508	5.4	6.5

[a] The comparison for unit weight of tellurium.
[b] The activity for the dehydrogenation of tetralin.
[c] The rate constants are defined in the text.
Source: Ref. 12.

is an irreversible consecutive first-order reaction. It has been proved that Eq. 14 can indeed be applied to the catalyses both by Te(liq.) and by Te(vapor) and, in addition, to the pyrolysis as well. Takahashi has thus determined the values of the rate constants $k_{1,liq}$, $k_{2,liq}$, $k_{1,vap}$, $k_{2,vap}$, $k_{1,p}$, and $k_{2,p}$ separately [the subscripts liq, vap, and p denote the catalysis by Te(liq.), the catalysis by Te(vapor), and the pyrolysis, respectively]. The effects of the reactor wall have been found to be negligible.

Presented in Table 2 are part of the rate constants obtained by Takahashi and Ogino [11,12] for the dehydrogenation of tetralin. As can be seen in the table, the intrinsic activity of Te(liq.) is several times larger than that of Te(vapor). This suggests that the dehydrogenation takes place on multiplet sites (each of them is composed of several catalyst atoms). The surface of

Kinetics 81

Te(liq.) is able to Provide the reacting tetralin with sufficient multiplet sites. On the other hand, the Te(vapor) provides the reactant with only a limited number of multiplet sites because only part of the particles in the vapor phase are polyatomic and the other particles are monoatomic [13]. The depression of contribution from Te(vapor) catalysis at a high temperature suggests that the fraction of polymeric Te particles in the vapor phase decreases on heating.

2. Decomposition of Benzyl Phenyl Ether

Studies on the decomposition of benzyl phenyl ether, a model compound of coal, have been carried out using Sn(liq.), Pb(liq.), Bi(liq.), In(liq.), Ga(liq.), Cd(liq.), and Zn(liq.) as catalysts [14,15]. As we have seen in an earlier chapter (Chapt. 3, VI, C), the catalysts promote the isomerization to o-benzylphenol. It must be pointed out, however, that the isomerization is only a small fraction of the decomposition and the reactions forming the random products (phenol, benzene, toluene, ethylbenzene, propylbenzene, and bibenzyl) predominate. This situation requires us to clarify the catalytic effects upon the reactions forming the random products. Thus, the results of benzyl phenyl ether decomposition have been analyzed kinetically.

Generally speaking, it is not easy to find a reasonable reaction scheme for a radical reaction because a complex product distribution hinders us from applying simple kinetic rules upon the experimental results. Indeed the reaction scheme shown in Fig. 2 was found after a number of trials [15]. As we can see below, this reaction scheme enables us to explain characteristic relations between the reaction products.

Under the assumptions that every step involved in the reaction scheme obeys irreversible first-order kinetics, it is possible to derive the following relations:

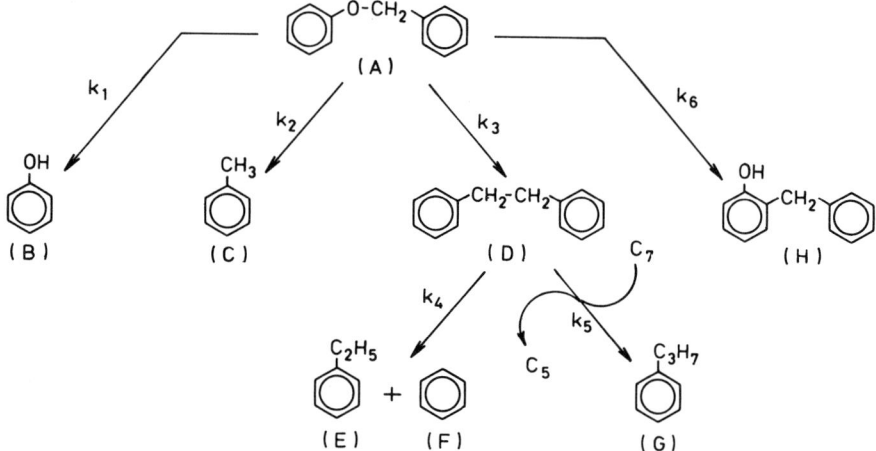

Fig. 2 A probable reaction scheme for the decomposition of benzyl phenyl ether [15].

$$\frac{X_B}{X} = \frac{1}{(1 + k_2/k_1 + k_3/k_1)} \tag{15}$$

$$\frac{X_C}{X} = \frac{k_2/k_1}{(1 + k_2/k_1 + k_3/k_1)} \tag{16}$$

$$\frac{X_D}{X} = \frac{k_3/k_1}{(1 + k_2/k_1 + k_3/k_1)} \tag{17}$$

$$\frac{X_E}{X_D^2} = \left(\frac{k_4}{2k_3}\right) \tag{18}$$

$$\frac{X_G}{X_D^2} = \left(\frac{k_5}{2k_3}\right) \tag{19}$$

where X is the total conversion of benzyl phenyl ether (A), X_i (i = B, C, D, E, F, and G) is the conversion of A to the product i, k_j (j = 1 - 5) is the rate constant for the jth step defined in Fig. 2.

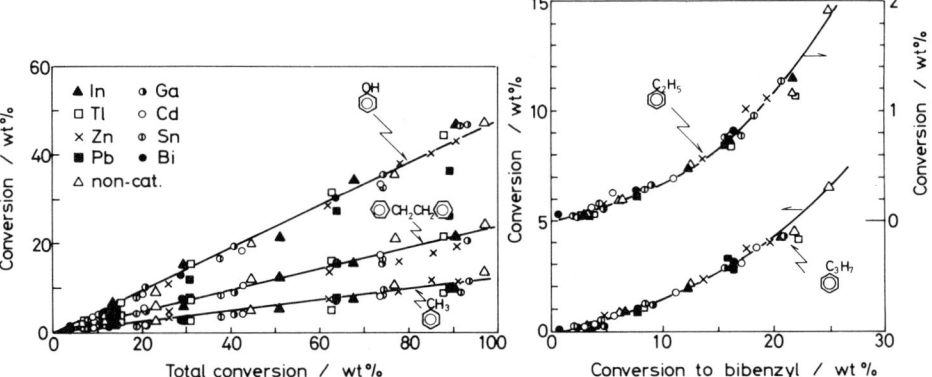

Fig. 3 Experimental results showing the validity of the reaction scheme for the decomposition of benzyl phenyl ether over various liquid metal catalysts [15].

The equations shown above predict that if the catalytic effects in every step forming the random products are little, each conversion ratio defined by the left-hand side of the equation would differ little from the respective values for the pyrolytic reaction, irrespective of the catalyst species used. As we can see in Fig. 3, the prediction mentioned above has been found to be valid in the light of the experimental results [15].

3. Coal Liquefaction

Various products such as gases (G), preasphaltenes (PA), asphaltenes (A), oils-1 (O1), and oils-2 (O2) are produced in a coal liquefaction reaction, and all products other than gases are mixtures of high-molecular-weight hydrocarbons. This situation creates difficulties in the kinetic study of the coal liquefaction and makes the kinetic scheme complicated [16]. Nevertheless Ozawa and his co-workers [17] have measured the yield of each product as a function of reaction time, to serve as the kinetic analysis.

With the reaction scheme shown in Fig. 4 and with the assumption that each step obeys the first-order irreversible kinetic

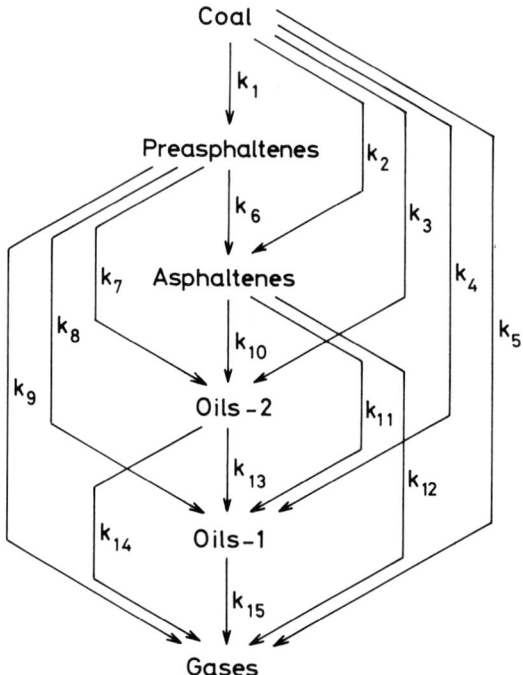

Fig. 4 The reaction scheme assumed for analyzing the kinetic data for the nonsolvent coal liquefaction over the Sn(liq.) catalyst [17].

law, the following rate equations have been derived

$$\frac{d\,PA}{dt} = k_1 C - (k_6 + k_7 + k_8 + k_9) PA \tag{20}$$

$$\frac{d\,A}{dt} = k_2 C + k_6 PA - (k_{10} + k_{11} + k_{12}) A \tag{21}$$

$$\frac{d\,O2}{dt} = k_3 C + k_7 PA + k_{10} A - (k_{13} + k_{14}) O2 \tag{22}$$

$$\frac{d\,O1}{dt} = k_4 C + k_8 PA + k_{11} A + k_{13} O2 - k_{15} O1 \tag{23}$$

$$\frac{d\,G}{dt} = k_5 C + k_9 PA + k_{12} A + k_{14} O2 + k_{15} O1 \tag{24}$$

$$C_0 = C + PA + A + O2 + O1 + G \tag{25}$$

TABLE 3 Rate Constants for the Elementary Steps of Coal
Liquefaction Reaction[a] (Shin-Yubari Coal)

		Catalyst		
		Sn(liq.)		
Rate constant (hr^{-1})		Set a	Set b	None
k_1	(C → PA)	0.118	0.114	0.035
k_2	(C → A)	0.132	0.144	0.034
k_3	(C → O2)	0	0	0
k_4	(C → O1)	0.064	0.059	0.035
k_5	(C → G)	0.017	0.018	0.020
k_6	(PA → A)	0.209	0.126	0
k_7	(PA → O2)	0	0	0
k_8	(PA → O1)	0	0.067	0
k_9	(PA → G)	0	0	0
k_{10}	(A → O2)	0.086	0.086	0.062
k_{11}	(A → O1)	0.032	0	0.068
k_{12}	(A → G)	0	0	0
k_{13}	(O2 → O1)	0	0	0
k_{14}	(O2 → G)	0	0	0
k_{15}	(O1 → G)	0	0	0

[a] Reaction temperature = 400°C; initial hydrogen pressure
= 8.0 MPa; catalyst/coal = 5/1 wt ratio.

Source: Ref. 17.

where t is the reaction time, C_0 is the initial coal concentration in the reaction system; and C, A, PA, O1, O2, and G are the respective concentrations of coal, asphaltenes, preasphaltenes, oils-1, oils-2, and gases (all in a weight basis). By applying the Marquardt's mathematical procedure [18], Ozawa and his coworkers [17] have analyzed the experimental results with rate equations shown above. The resulting rate constants are given in Table 3. As we can see in the table, two sets of the rate

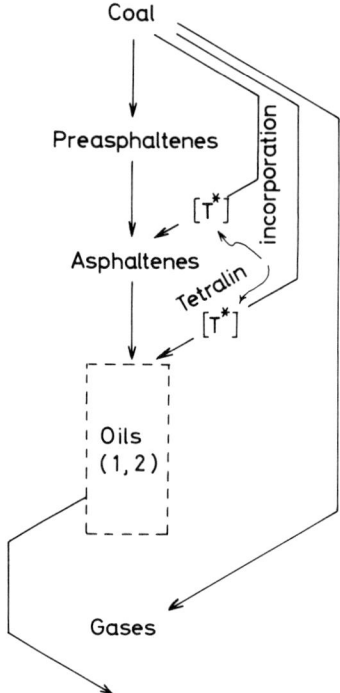

Fig. 5 A probable reaction scheme for the coal liquefaction in the presence of Sn(liq.) as a catalyst and tetralin as a hydrogen donor solvent [19].

constants (set a and set b) are equally capable of representing the experimental results. Owing to the complexity of the reaction, it is hard to determine which set of the rate constants is better. However, we can see in the table that steps 1 and 2 and in particular step 6 have been promoted by the Sn(liq.) catalyst, irrespective of our choice of a or b. This means that Sn(liq.) exhibits a particularly high catalytic activity for step 6, step PA to A. This is in sharp contrast to the low catalytic activity of Sn(liq.) for steps 3, 7, and 8, each forming oil fractions. The kinetic data shown above and the structural parameters of the products (Table 7, Chapt. 3) enable us to make a crude reaction model, as shown in the preceding chapter (Chapt. 3, VI).

Kinetic treatments similar to those mentioned above can be made with the liquefaction data obtained under the coexistence of Sn(liq.) and tetralin. The resulting reaction scheme is shown in Fig. 5 [19]. An important point to be mentioned here is that the respective rate equations for the two steps (coal → oils-1, -2; coal → asphaltenes) have to include a first-order term with respect to the tetralin concentration.

III. STEREOCHEMISTRY

Certain hydrogen transfer reactions between alcohols and ketones over the liquid metal catalyst exhibit stereochemical characteristics. For instance, Miyamoto and Ogino [20] have reported that an axial attack of an alcohol molecule to methylcyclohexanone molecule is somewhat preferential to the equatorial attack over the In(liq.) catalyst. This is not unlikely because the hydrogen transfer reaction over the liquid metal catalyst is considered to take place as a consequence of a surface bimolecular collision that probably passes through the following transition state [2,20]:

$$\begin{array}{c} R_1 \\ {\diagdown} \\ R_2 \end{array} \begin{array}{c} C \cdots H \cdots O \\ \vert \vert \vert \\ O \cdots H \cdots C \end{array} \begin{array}{c} \\ \\ {\diagup}{\diagdown} \\ R_3 R_4 \end{array}$$

The formation of the transition state shown above is probably restricted by the molecular structure of alcohol and that of ketone, and hence the rate of reaction is considered to be sensitive to the steric hindrance. This view is supported by the experimental results shown in Fig. 6 [21]. It can be seen that the conversion of the hydrogen transfer from menthol to the

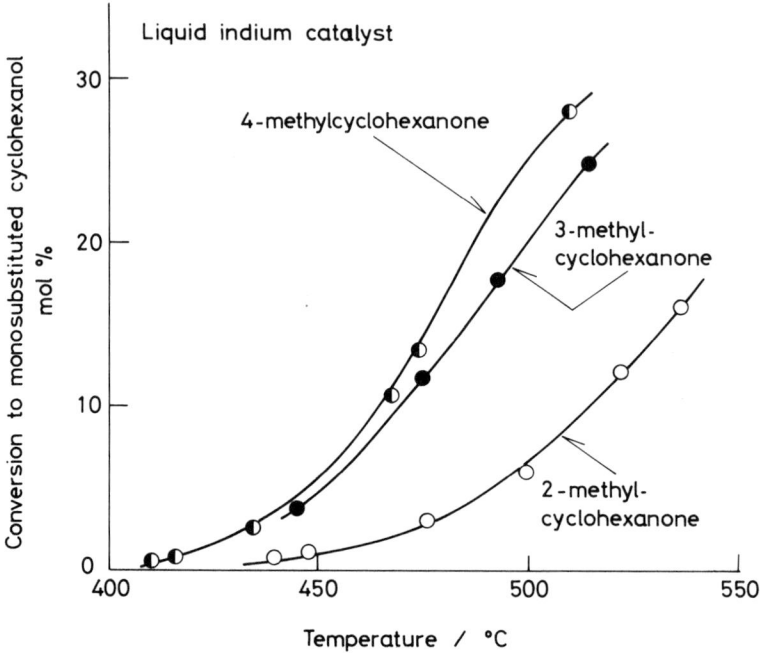

Fig. 6 The hydrogen-accepting ability of methylcyclohexanone from menthol as a function of the position of the methyl group on the benzene ring [21].

monosubstituted cyclohexanone is a function of the position of the substituent, i.e., 2-position < 3-position < 4-position;

(27)

In connection with the problem of the steric hindrance, the hydrogen transfer reaction between optically active compounds deserves special attention. The specificity of this kind of reaction can be best understood with the aid of Fig. 7 [21]. This figure schematically represents the transition state of two reactions: One is the reaction between ℓ-menthone and d-fenchol and

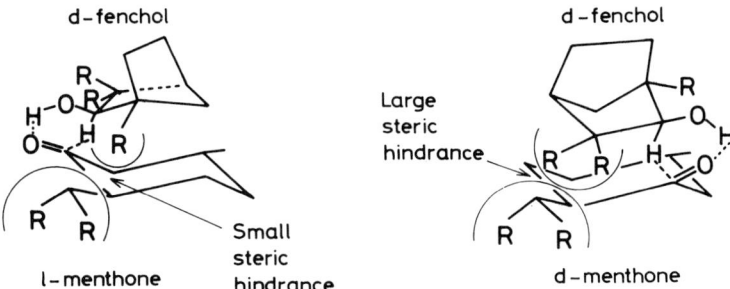

Fig. 7 Transition state models for the ℓ-menthone/d-fenchol hydrogen transfer system and for the d-menthone/d-fenchol hydrogen transfer system [21].

the other is the reaction between d-menthone and d-fenchol. It is evident that the steric hindrance for the former reaction is smaller than that of the latter reaction. Therefore we can expect that the conversion of the former reaction would be larger than that of the latter reaction. The experimental results shown in Fig. 8a support the view mentioned above. We can see in Fig. 8b that the difference between the conversion of the ℓ-isomer and that of the d-isomer can be controlled by changing the reaction conditions.

A more elegant kinetic study on the enantiomer differentiating reaction between menthone and fenchol over In(liq.) has been published [22]. It is reported that the initial rate r_0 is expressed by

$$r_{0,j} = k_j p_A \qquad (28)$$

where the subscript j denotes ℓ or d, k_j denotes the rate constant for the j-isomer of menthone, and p_A is the partial pressure of d-fenchol.

The Arrhenius plots shown in Fig. 9 clearly show that $k_\ell > k_d$. From the analysis of the rate equation, the following information has been derived: (a) The surface of the In(liq.) catalyst is almost completely covered by adsorbed menthone

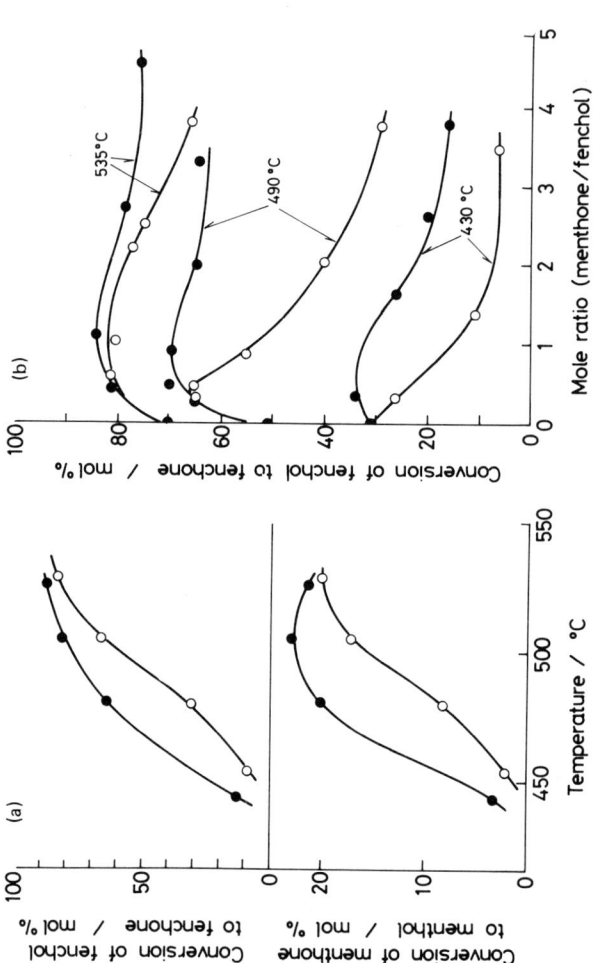

Fig. 8 Influences of the reaction conditions upon the enantiomer-differentiation observed in the menthone/fenchol hydrogen transfer system: (●) d-fenchol ~ ℓ-menthone, (○) d-fenchol ~ $d\ell$-menthone. (a) Influence of the reaction temperature; (b) influence of the feed composition [21].

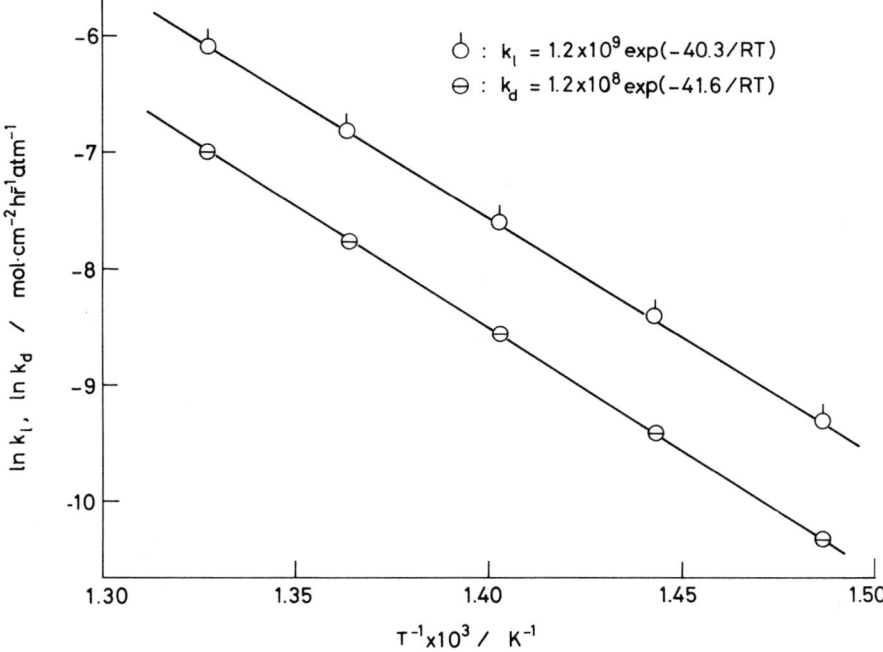

Fig. 9 Arrhenius plots for two kinds of fenchol/menthone hydrogen transfer reactions: k_ℓ, the rate constant for the d-fenchol/ℓ-menthone system; k_d, the rate constant for the d-fenchol/d-menthone system [22].

molecules; (b) the chemisorption of fenchol molecules upon the adsorbed menthone molecules limits the reaction rate.

There are many possible reactions similar to those mentioned above, e.g., a borneol-menthone system and a fenchol-camphor system [4,21]. Experimental results for these reactions can also be explained in terms of the steric hindrance. Furthermore, even in a seemingly simple alcohol dehydrogenation, the ketone produced can accept hydrogen from its parent alcohol existing in the reaction system. Thus, when the hydrogen transfer reaction is capable of forming several steric isomers, the complexity of the reaction scheme multiplies. For instance, Saito and Ogino [4,23]

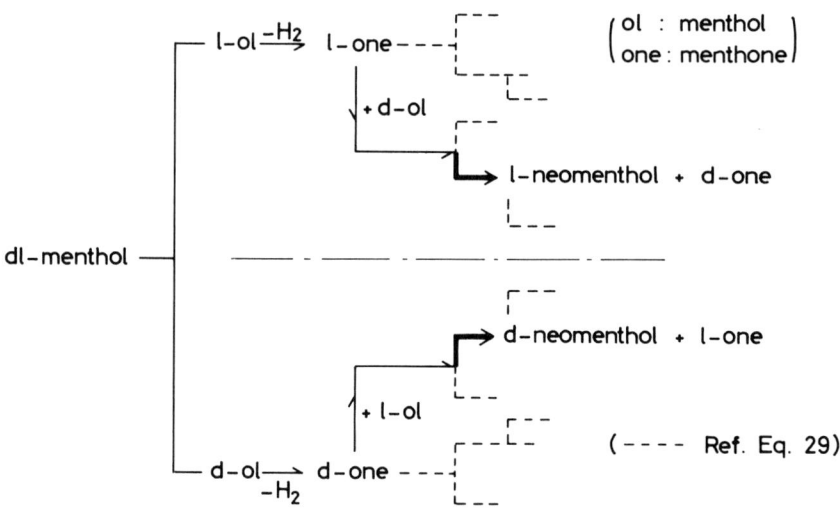

Fig. 10 The fine structure of the reaction pathway proposed for the reaction of $d\ell$-menthol over the liquid metal catalyst [4,23].

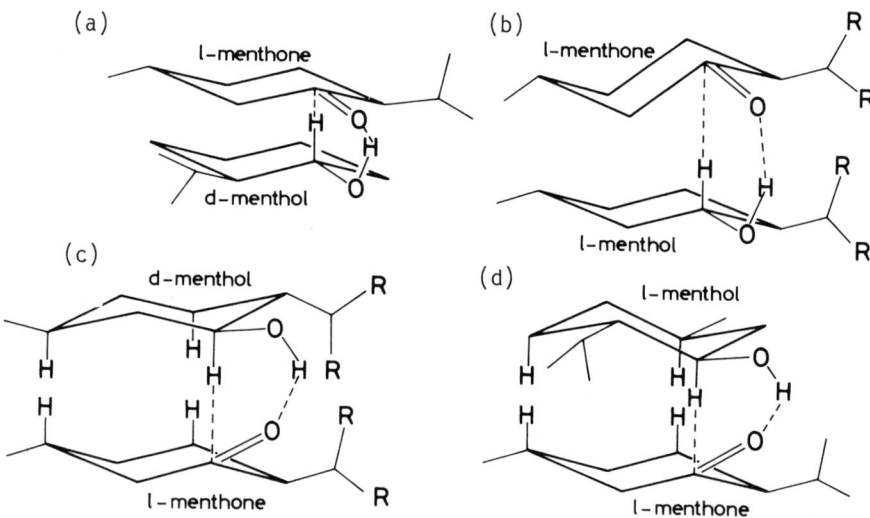

Fig. 11 Transition state models with different degrees of the steric hindrance [4,23].

TABLE 4 Experimental Results of Reaction between ℓ-Menthol and dℓ-Menthol over Liquid Metal Catalysts

Liquid metal	Reaction temperature (°C)	dℓ-Menthol Conversion (mol%) to		ℓ-Menthol Conversion (mol%) to	
		menthone	neomenthol	menthone	neomenthol
Tl	440	5.3	0.8	3.7	0.4
	479	13.8	3.3	10.5	2.2
	500	19.2	5.7	13.8	2.8
	524	26.9	8.2	21.2	4.9
In	445	3.8	0.5	4.4	0.3
	476	12.9	2.5	6.0	0.6
	498	18.1	4.4	9.6	1.3
	523	(27.0)[a]	(7.5)[a]	19.8	5.0

[a]Data obtained at 530°C.
Source: Refs. 4 and 23.

have proposed the following reaction scheme for the dehydrogenation of menthol over the In(liq.) catalyst or Tl(liq.) catalyst;

$$\text{Menthol} \xrightarrow{-H_2} \text{Menthone} \xrightarrow{+\text{Menthol}} \begin{array}{l} \to \text{Menthol} + \text{Menthone} \\ \to \text{Neomenthol} + \text{Menthone} \\ \quad \downarrow -H_2 \\ \quad \to \text{Menthone} \end{array} \quad (29)$$

It must be noted that this reaction scheme is only applicable to the reaction of either ℓ-isomer or d-isomer.

When we use dℓ-menthol as a reactant, the reaction scheme becomes more complex [23], and there are new reaction pathways where the steric hindrances are smaller than those for other steps. The new steps are depicted in Fig. 10 by bold solid lines. The transition-state models shown in Fig. 11 clearly show that the steric hindrances for the new steps mentioned above (model a in Fig. 11) are the smallest [23]. On the basis of the reasons mentioned above, the $d\ell$-menthol reacts faster than d-menthol or ℓ-menthol, as we can see in Table 4 [4,23].

IV. PROBLEMS IN CATALYSIS BY LIQUID ALLOY

A. Catalytic Activity and Alloy Composition

Measurements of catalytic activities of binary liquid alloys as a function of the alloy composition occasionally bring about unexpected results. For instance, as we can see in Fig. 12, an incorporation of ~10 atom% of Zn into an inactive liquid metal

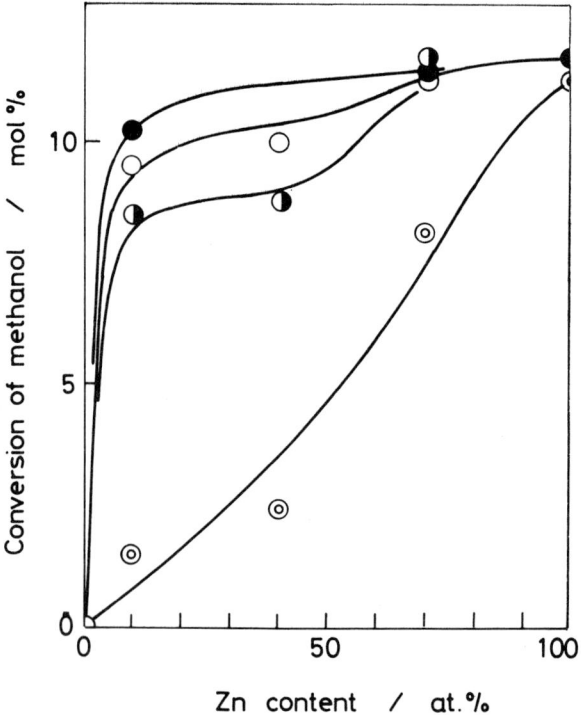

Fig. 12 Methanol decomposition activities of several zinc based binary liquid alloy catalysts at 540°C [4]. (●) Zn-Bi, (o) Zn-Pb, (◐) Zn-Sn, (◎) Zn-Cd.

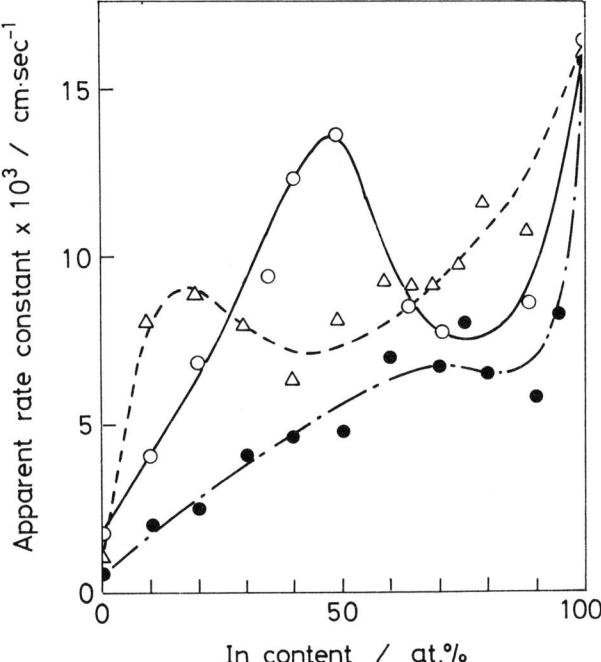

Fig. 13 2-Butanol dehydrogenation activities of some binary liquid alloy catalysts at 500°C [4]: (●) In-Bi, (○) In-Sn, (△) In-Pb.

such as Sn, Pb, or Bi causes a marked elevation of the methanol decomposition activity. On the other hand, no such clear activity change is brought about by the incorporation of Zn into Cd(liq.) [4].

A much more interesting relation between the catalytic activity and the alloy composition has been found for the binary liquid alloy catalyst containing In as an active component [4]. As we can see in Fig. 13, every curve representing the relation between the catalytic activity and the alloy composition exhibits curious shapes with maximum and minimum. Furthermore, changes in the reactant bring about the shape change of the activity-composition curve, as we can see in Fig. 14 [4]. Except for the

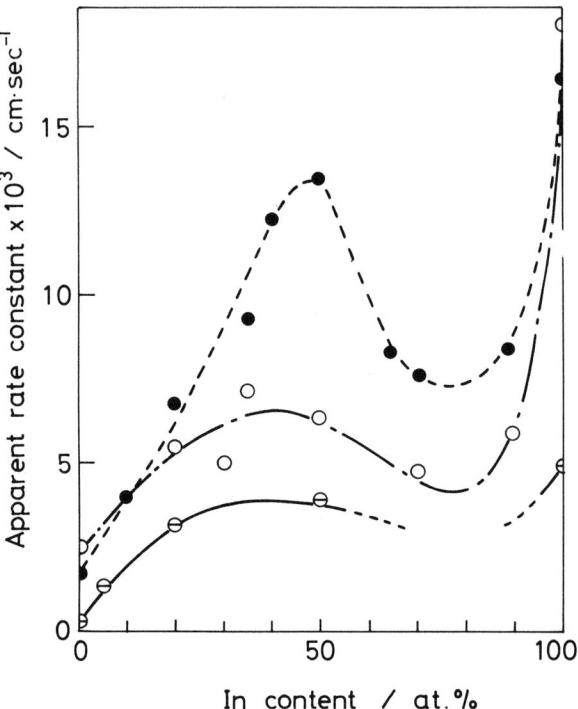

Fig. 14 Catalytic effects of the In-Sn liquid alloy upon the dehydrogenation of three different alcohols [4]: (●) 2-butanol, 500°C, (o) 2-propanol, 500°C, (⊖) ethanol, 505°C.

case of the Cd-Zn system, all of the experimental results mentioned above strongly suggest some structural changes of catalysts.

B. Atomic Distribution in Liquid Alloy

X-ray scattering is a powerful tool for studying the bulk structure of a liquid alloy, although an electron diffraction is also available [24,25]. However we need some special equipment and techniques to perform the X-ray scattering study of the liquid metal or liquid alloy [26-29]. For instance, as illustrated in Fig. 15, we have to use an apparatus in which the specimen is

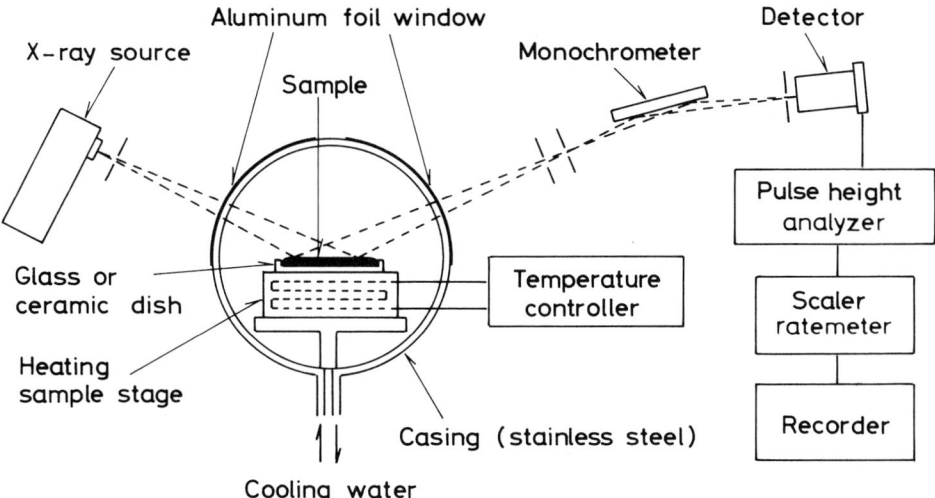

Fig. 15 An arrangement of X-ray scattering units for the specimen in a liquid state at a high temperature and under an inert gas atmosphere [4].

held in a horizontal position. In addition, the specimen has to be held at a high temperature and protected from oxidation. A step-scanning unit equipped with an automatic digital recorder facilitates accurate measurements of intensities of the X-ray scattered by the specimen. The radiation intensity thus obtained is corrected and normalized [30,31] to obtain an intrinsic intensity, which is exemplified in Fig. 16 [4].

Information about the structure of liquid alloy can be obtained by analyzing the intrinsic intensity with conventional procedures [28,30,32]: The theoretical basis of the analysis is well established [33], and a high-speed digital computer greatly assists numerical treatments. The total radial distribution function $4\pi r^2 \rho(r)$ representing the probability $\rho(r)$ of finding other atoms located at a distance r from any specified atom is thus obtained. In addition, with appropriate assumptions, the partial radial distribution function $4\pi r^2 \rho_{A-B}(r)/C_B$ can also be

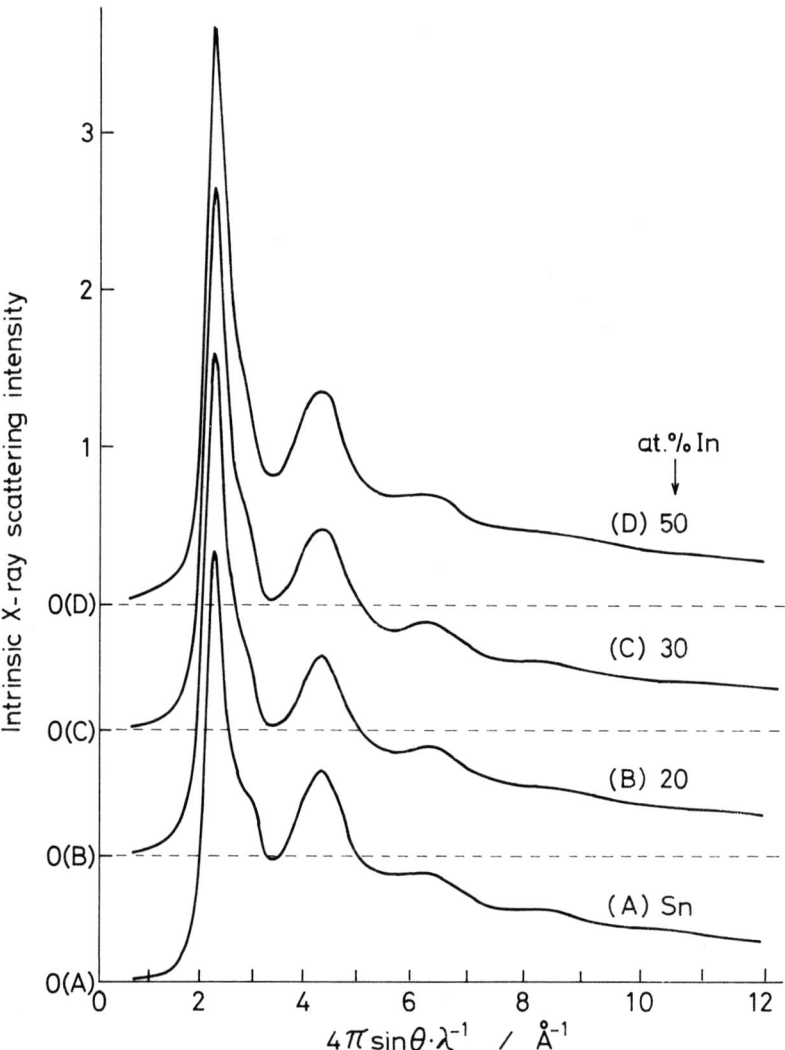

Fig. 16 Intrinsic X-ray scattering intensities at 550°C for several In-Sn liquid alloys with different compositions as a function of the diffraction vector $4\pi \sin\theta/\lambda$, where 2θ is the scattering angle and λ is the wavelength of the X-ray used [4].

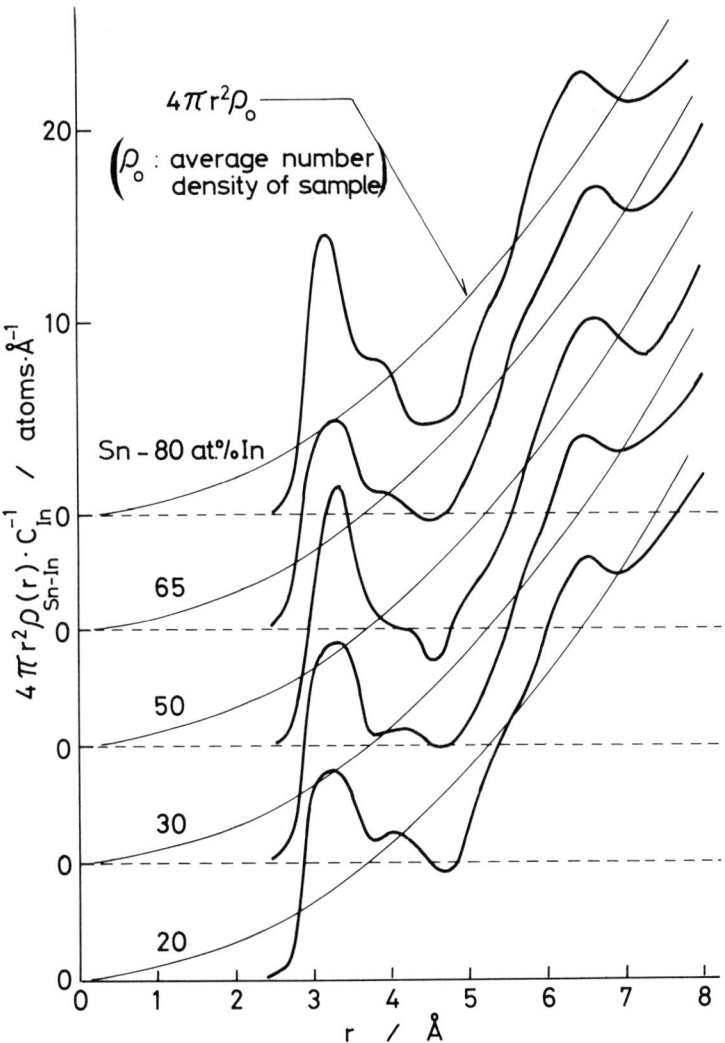

Fig. 17 Partial radial distribution functions at 550°C for several In-Sn liquid alloys with different compositions as a function of the atomic distance r from any Sn atoms in the alloy [4].

TABLE 5 Structural Data for the In-Sn Liquid Alloy[a]

In content (atom%)	Nearest neighbor distance (Å)	Partial coordination number[b]			
		In-In	Sn-In	In-Sn	Sn-Sn
0	3.2	-	-	-	8.5
20	3.2	1.9	1.9	7.5	6.8
30	3.3	2.9	2.9	6.7	6.0
50	3.3	4.9	4.8	4.8	4.3
65	3.2	6.3	6.1	3.3	3.0
80	3.2	7.8	7.8	2.0	1.7
100	3.2	9.7	-	-	-

[a] at 550°C.
[b] The number of the nearest neighbor atoms around the underlined atom.

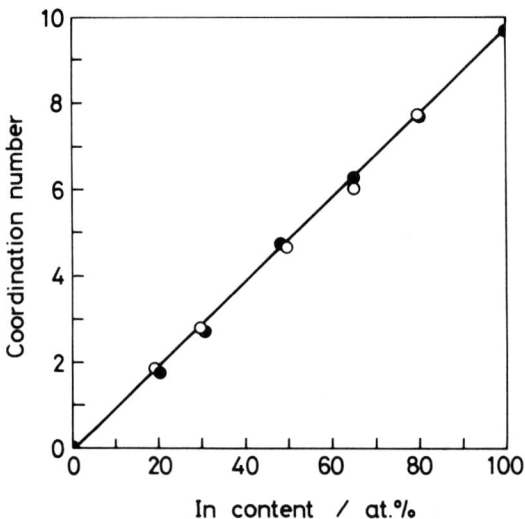

Fig. 18 The number of In (●) atoms existing at the nearest neighbor distance from any central atom (atoms of the element underlined in the figure can be the central atom) taken in the In-Sn (o) liquid alloy at 550°C [4].

obtained: ρ_{A-B} is the probability of finding B atoms locating at a distance r from a specified A atom in a binary liquid alloy A-B, and C_B is the concentration of B component in the bulk of the alloy. Illustrated in Fig. 17 [4] are the partial distribution functions of the In-Sn liquid alloys with different compositions.

From the radial distribution function mentioned above, it is possible to find the nearest neighbor distance, coordination numbers, and the composition in the vicinity of a central atom, as we can see in Table 5 [4]. Part of the data shown in this table are graphically presented in Fig. 18, in order to facilitate the comparison with the catalytic data shown in the previous figures (Figs. 13, 14). It is evident that there is no direct correlation between the catalytic activity of the In-Sn liquid alloy and its bulk structure, which can be regarded as a typical random mixture of Sn and In atoms. Results similar to this have been obtained [4] by comparing X-ray scattering results for In-Pb liquid alloy [4] as well as for the In-Bi liquid alloy [34] with the catalytic activities (Fig. 13).

C. Surface Composition

1. *Surface Tension and Surface Composition*

In the discussion of surface catalysis, information about the composition of the surface is indispensable. Among several methods that are available for the surface analysis of the liquid alloy, the surface tension method is a traditional one. The purpose of this section is to present information relevant to the surface catalysis and hence general problems in the surface tension studies are not included herein. Wider information about the surface tension will be presented in the next chapter (Chapt. 5).

There are various methods for measuring the surface tension [35] but the maximum bubble pressure method looks most appropriate for the measurement of the surface tension of a catalytically

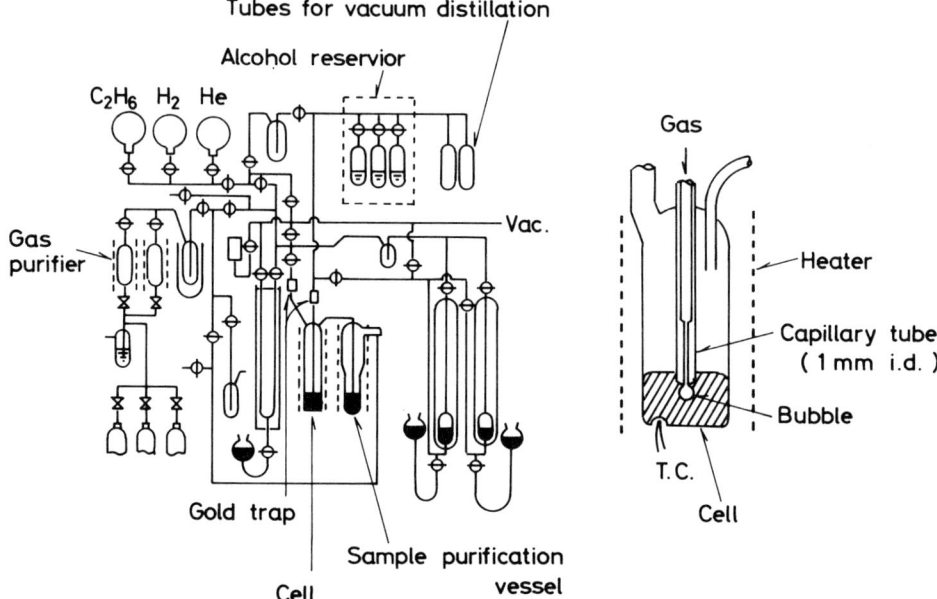

Fig. 19 A maximum bubble pressure apparatus for the measurement of surface tension of liquid metals and liquid alloys [4,36].

active liquid alloy. This method enables us to make the measurement at a high temperature and under any desired gas (or vapor) atmosphere. Therefore information about the gas (vapor) adsorption onto the sample liquid can also be obtained. The apparatus shown in Fig. 19 [4,36] serves for the measurement mentioned above.

Shown in Fig. 20 [4,36] are the surface tension data for the In-Sn liquid alloy. It is evident that hydrogen is not adsorbed, whereas 2-butanol vapor is adsorbed by the liquid alloy. Thus, it is expected that accumulation of the surface tension data would bring about useful information about the surface catalysis of the liquid alloy. Indeed, as we can see below,

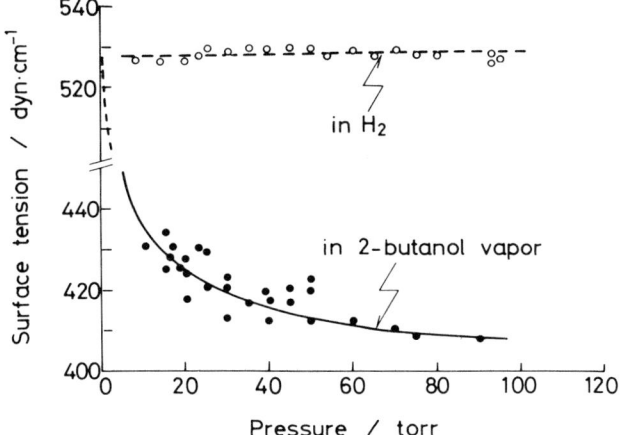

Fig. 20 Effects of atmosphere upon the surface tension of the In-Sn liquid alloy at 350°C (65 atom% of In) [4,36]: (●) in 2-butanol vapor, (o) in H_2.

analyses of the surface tension of the In-Sn liquid alloy yield interesting results.

The surface composition of the In-Sn liquid alloy can be obtained by putting the surface tension data into the following equation [35]

$$-d\gamma = \hat{\Gamma}_1 kT \left[\frac{1 + \partial \ln \phi_1 / \partial \ln x_1^b}{x_1^b(1 - x_1^b)} \right] dx_1^b + \Gamma kT d \ln p \qquad (30)$$

where γ is the surface tension, $\hat{\Gamma}_1$ is the surface excess of the alloy component specified as 1, k is Boltzmann's constant, T is the absolute temperature, x_1^b is the atomic fraction of 1 in the bulk of the alloy, ϕ_1 is the activity coefficient of the component 1, Γ is the amount of adsorption of gas (vapor) employed to make the bubble, and p is the pressure of the gas (vapor). Thermodynamic data necessary for the evaluation of the activity coefficient ϕ_1 can be obtained from literature [37].

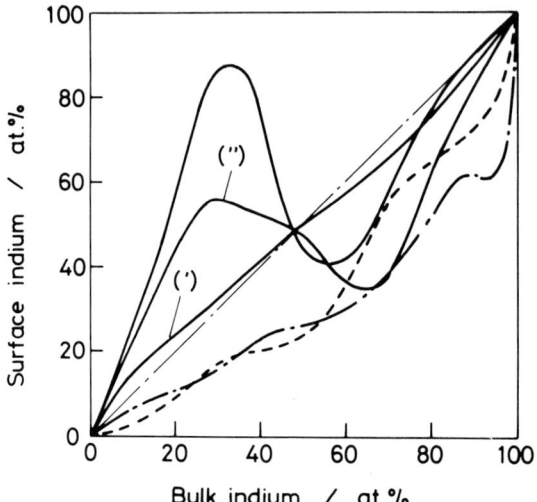

Fig. 21 The surface composition as a function of the bulk composition of the liquid alloy: (———) In-Sn under 100 Torr of 2-butanol vapor at 350°C, (———') In-Sn in vacuum at 350°C, (———") In-Sn in vacuum at 250°C, (— ·· — · —) In-Bi in vacuum at 450°C, (-----) In-Pb in vacuum at 500°C [4].

The result of this calculation is shown in Fig. 21 (———) [4,36]. It must be pointed out that, under the atmosphere of 2-butanol vapor, the relation between the surface composition and the bulk composition exhibits a curve similar to that representing the relation between the catalytic activity and the bulk composition (Fig. 13). Unfortunately, however, there is no exact proportionality between the catalytic activity and the surface concentration of In. This implies that the true relation between the catalytic activity and the surface composition is more complex than that expected initially.

From the reinvestigation of the curves ———, ———', and ———" shown in Fig. 21, it turns out that the surface composition x_{In}^{s}/x_{Sn}^{s} at $x_{In}^{b}/x_{Sn}^{b} = 1$ is held at a constant value, i.e.,

$x_{In}^S/x_{Sn}^S = 1$, irrespective of the temperature and the atmosphere of the measurement of the surface tension. Furthermore, the deviation of the actual surface concentration of In from the ideal one changes the sign from positive to negative at $x_{In}^b/x_{Sn}^b = 1$. These facts give an impression that an equimolar surface complex of In and Sn is formed at $x_{In}^b/x_{Sn}^b = 1$. Unfortunately, there is no direct evidence for the existence of the surface complex, but the following fact again gives us an impression similar to that mentioned above.

As we can see in Fig. 22 [4,38], the adsorption equilibrium constant K_A maximizes at $x_{In}^b/x_{Sn}^b = 1$. The adsorption equilibrium

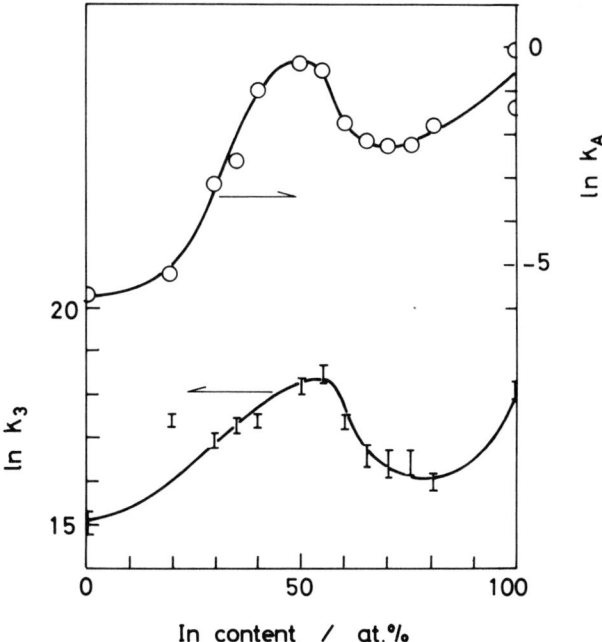

Fig. 22 A comparison between the equilibrium constant K_A for the 2-butanol vapor adsorption onto the In-Sn liquid alloy at 350°C and the rate constant k_3 for the surface unimolecular decomposition step in the 2-butanol dehydrogenation over the In-Sn liquid alloy at 465°C [4,38].

constant K_A is related to the surface tension by

$$\gamma_0 - \gamma = N_s k T K_A \int_0^p \frac{dp}{(1 + K_A p)} \tag{31}$$

where γ_0 is the surface tension of the liquid alloy in an inert atmosphere and N_s is the number of atoms per unit of surface area. It is evident from Eq. 31 that the maximum of K_A corresponds to the maximum of the free energy of adsorption, $(\gamma_0 - \gamma)/N_s$. Therefore we can say that the surface with the composition of $x_{In}^s/x_{Sn}^s = 1$ is most favorable for the adsorption of 2-butanol. In this connection, Fig. 22 [4,38] deserves special attention. It can be seen in this figure that a good parallelism exists between K_A and k_3 (the rate constant of the surface unimolecular decomposition of the activated admolecules; see Eq. 6, in II of this chapter). Thus, it turns out that the surface with most favorable sites for 2-butanol adsorption catalyzes most effectively the dehydrogenation of the adsorbed 2-butanol.

The surface tension data published by Kovalchuk [39] and the activity coefficients reported by Ryabov [40] enable us to evaluate the surface composition of In-Bi liquid alloy. Similarly, the surface composition of the In-Pb liquid alloy can be evaluated from the surface tension data of Hoar [41] and from the thermodynamic data of Hultgren [37]. The results of calculation for both liquid alloys are shown in Fig. 21 (— - — - and -------). Although there is a crude parallelism between the surface composition (Fig. 21 [— - — -]) and the catalytic activity of the In-Bi liquid alloy (Fig. 13 [●]), little parallelism exists between the surface composition and the catalytic activity of the In-Pb liquid alloy (Fig. 21 [-----] and Fig. 13 [△]). Surface tension measurements in the presence of 2-butanol vapor appear to be necessary for these two alloy systems, to make further clarification.

2. Auger Electron Spectroscopy

A wider review on the modern techniques of the surface analysis of the liquid alloy is presented in Chapter 7. Therefore, the description in this section concentrates upon the methods and results of Auger electron spectroscopy (AES) surface analysis of the In-Sn liquid alloy for which the catalytic activity is already known.

An AES apparatus illustrated in Fig. 23 [42] has been used in studying the surface composition of the In-Sn liquid alloy. Because the sample has to be kept in a liquid state at a high temperature, the design of the AES apparatus is somewhat restricted. The apparatus shown in the figure holds the common axis for the CMA and for the electron gun vertically, in order to analyze the chemical composition of the free surface of the liquid alloy.

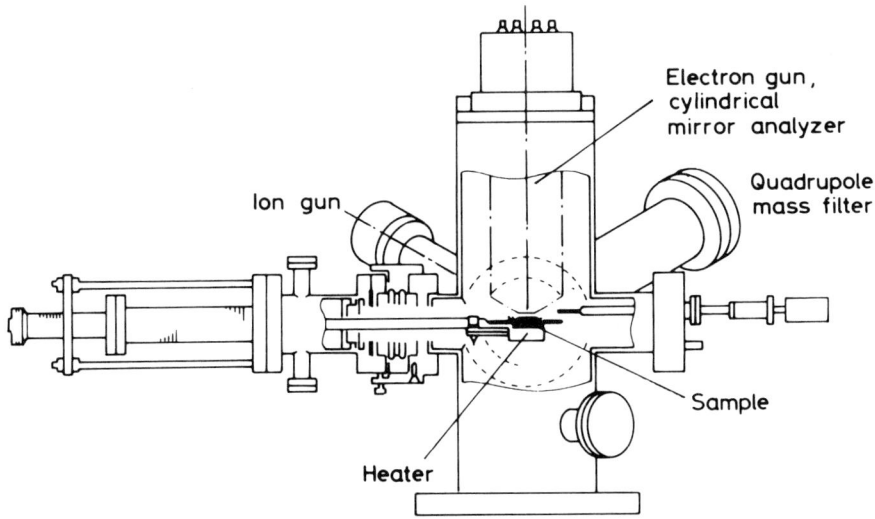

Fig. 23 A schematic drawing of the Auger electron spectroscopic apparatus designed for studying the surface composition of the liquid alloy [42].

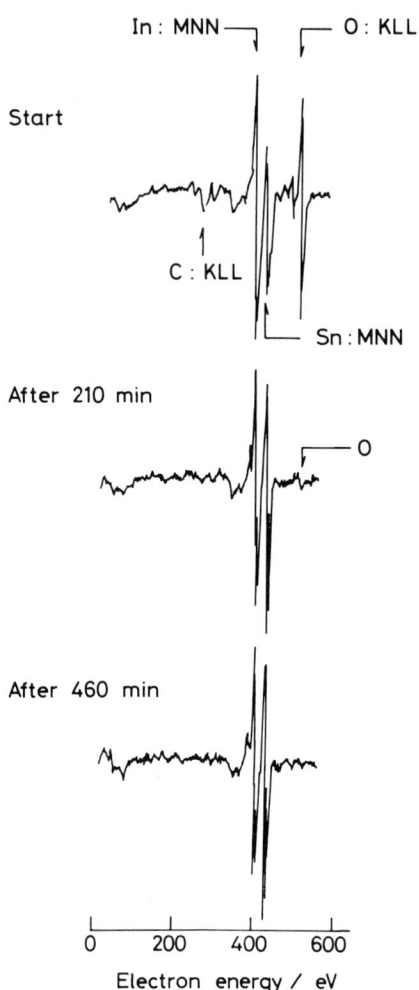

Fig. 24 Changes in the AES (Auger electron spectroscopy) signal of the In-Sn liquid alloy containing 40 atom% of In during the Ar^+ bombardment in the presence of hydrogen at 160°C [42].

Although an evacuation to ~10^{-9} Torr of pressure is not so difficult when the sample is in a solid state at a low temperature, the evacuation becomes difficult when the temperature of the sample is elevated above the melting point of the sample alloy; diffusion of gases from interior of the sample continues, and the surface of the sample is thus contaminated. The surface contamination by oxygen is particularly severe and hard to remove by a conventional Ar ion-sputtering technique (Ar pressure = 10^{-4} Torr, voltage 2 kV, emission current 25 mA).

The cleaning of the sample surface is achieved by an Ar ion sputtering in the presence of hydrogen (10^{-5}-10^{-6} Torr) in the sample chamber. Changes in the AES signal during the sputtering are illustrated in Fig. 24 [42]. We can see in this figure that the surface of the sample contains abundant oxygen before the sputtering and about one-quarter of the surface oxygen survives the exposure to the sputtering conditions during a 2-hr period. Thus the clean surface is obtained after ~9 hr of the sputtering.

The temperature dependence of the AES signal intensity ratio I_{In}/I_{Sn} is illustrated in Fig. 25 [42]. A larger intensity ratio

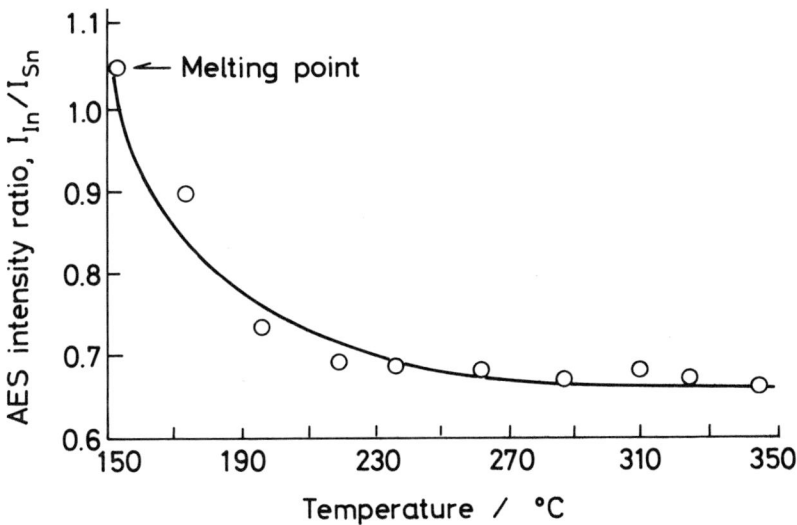

Fig. 25 The AES intensity ratio I_{In}/I_{Sn} as a function of the temperature for the In-Sn liquid alloy containing 40 atom% of In [42].

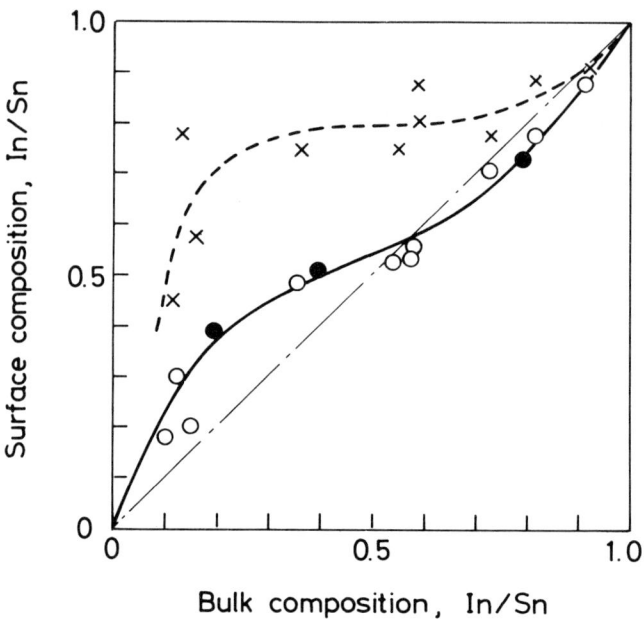

Fig. 26 The surface compositions of In-Sn alloys with different bulk compositions and in different states [42]: (-----) solids with O/(Sn + In) = 0.5-0.9, (●) liquid, and (○) solid at melting point.

at the melting point indicates a greater degree of the surface segregation of In at this temperature. The intensity ratio rapidly decreases on increasing the sample temperature and reaches a constant value at high temperatures. This means that the surface segregation of In decreases on increasing the temperature and disappears at high temperatures.

The relation between the surface composition and the bulk composition of the In-Sn liquid alloy is shown in Fig. 26 [42]. We can see in this figure that the surface segregation of In is larger for the liquid alloy with an In content less than 50 atom%. It must be noted, however, that the surface segregation is only appreciable in the vicinity of the melting point.

It is interesting that, as we can see in Fig. 26, the surface composition of the In-Sn solid alloy at the melting point is almost identical with that of the liquid alloy. This enables us to expect that the AES analysis data for solid alloy might be available to estimate an approximate surface composition of the liquid alloy, at least at the melting point. In this connection, the solid-state surface composition of the In-Sn alloy with significant amount of oxygen on the surface is interesting. As we can see in Fig. 26 (dashed line), the surface segregation of In on the oxygen-contaminated surface is much more marked than that on the clean surface. It may be inferred that the situation will be the same in the liquid state.

The results of these AES studies encourage us in planning future studies. Studies on the interaction between any reactive molecules and the liquid metal surface with a controlled oxygen content appear interesting. With respect to this, the molecular beam techniques reported by Balooch et al [43] (Chapt. 3, VII, B) are of great significance.

V. ELECTRONIC ASPECT OF CATALYSIS

A. Qualitative Results

A few experimental results strongly suggest that the electron-donating property of a liquid metal is closely related to the catalytic activity of this metal. For instance, as shown in Fig. 27 [44], every metal that catalyzes the dehydrogenation of alcohol (Chapt. 3, II) has the smallest first ionization potential among the elements of the same group, with the exception of zinc. Furthermore, any one of the liquid alloys that are active for the dehydrogenation of hydrocarbons contains either Na or K as an active component (Chapt. 3, IV). The smallest ionization potentials of these two elements are also evident in the figure. Addison [45] has also reported that the reactivity of the surface of an alkali metal in the liquid state strongly depends on its electron-donating ability.

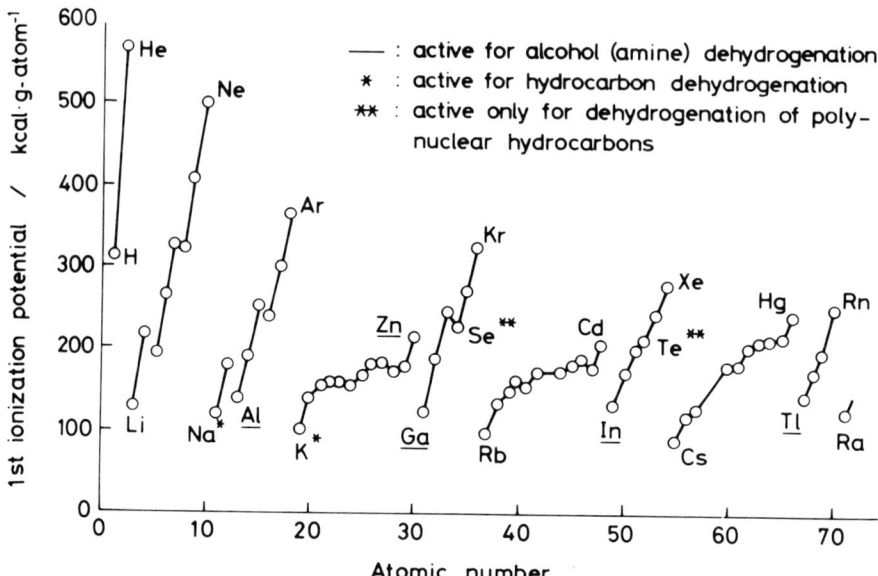

Fig. 27 Positions of the catalytically active elements on the diagram, showing a periodic relation between the first ionization potential and the atomic number [44].

The importance of the electron-donating property of the liquid metal can also be deduced from the comparison between the work function and the catalytic activity. For instance, it is possible to show that almost all metals having a small work function in their solid states are catalytically active in their molten states [46]. This fact appears to suggest that the work function of every metal changes little upon melting, and hence the work function of the catalytically active metal is small. According to Norris [47], the work function of a metal changes little on melting.

The catalytic activities of liquid alloys can also be related to the work function. As we can see in Fig. 28 [48], the catalytic activity of the binary liquid alloy containing sodium can be correlated with the hypothetical work function W_H which

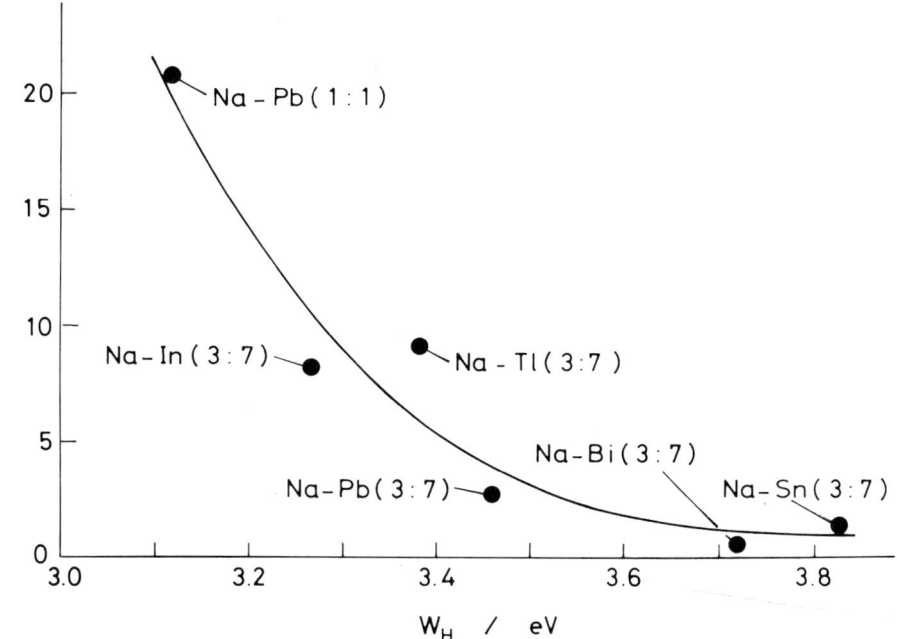

Fig. 28 The relation between the hypothetical work function W_H and the ethylbenzene dehydrogenating activity for binary liquid alloys containing Na at 550°C [48].

is defined by

$$W_H = x_{Na}W_{Na} + (1 - x_{Na})W_M \qquad (32)$$

where x_{Na} is the atomic fraction of sodium, W_{Na} is the work function of sodium in the solid state, and W_M is the work function of the metal M mixed with sodium and is approximated by the value in the solid state. Obviously, the hypothetical work function is not the true work function; nevertheless, the good correlation between the catalytic activity and W_H appears worthwhile. These results show the importance of studying the work function of liquid metals or liquid alloys.

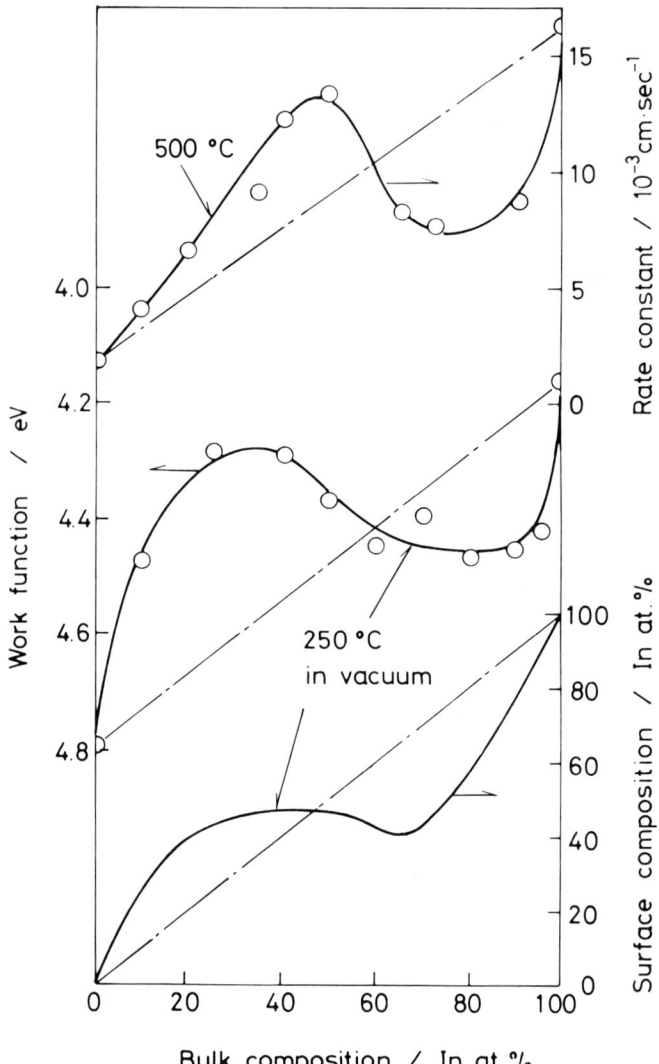

Fig. 29 Parallelisms among the work function, the 2-butanol dehydrogenating activity, and the surface concentration of In for the In-Sn liquid alloy system [51,52].

Electronic Aspect of Catalysis 115

Information about the work function of liquid metal is still scarce [47,49,50] and that of liquid alloys is not available. Thus, the data obtained in the author's laboratory are presented in Fig. 29 [51,52]. This figure represents the work function of the In-Sn liquid alloy as a function of the alloy composition. The catalytic activity and the surface composition determined by the surface tension method are also shown in the same figure. The respective curves representing the catalytic activity, surface composition, and work function have analogous shapes. This result suggests that the three quantities mentioned above are intimately interrelated. However, it must be noted that, as we have seen in the preceding section (Chapt. 4, IV, C, 2), considerable amounts of oxygen exist on the alloy surface. Work function studies with samples carrying controlled amounts of oxygen appear indispensable for further progress.

B. Semi-Quantitative Results

1. *Electron Delocalization*

Taking into account the qualitative results mentioned above and experimental results mentioned in the previous chapter (Chapt. 3, II), Miyamoto et al [2,46] have made a quantum chemical study on the dehydrogenation of alcohol over the liquid metal catalyst. The model employed by him is shown in Fig. 30 [2,46], and the basic equation is as follows:

$$\Delta E = \sum_{s=1}^{2} \left[2 \sum_{j}^{unocc} \frac{c_{j,s}^2}{\varepsilon_j - \varepsilon_F} + 2 \sum_{j}^{occ} \frac{c_{j,s}^2}{\varepsilon_F - \varepsilon_j} \right]$$
$$\times \sum_{k,\varepsilon_k = \varepsilon_F} \left| \int X_s H' \varphi_k d\tau \right|^2 \quad (33)$$

where ΔE is the electron delocalization energy; $C_{j,s}$ (s = 1, 2, which specify the two hydrogen atoms shown in Fig. 30) is the LCAO-MO coefficient of the sth hydrogen atomic orbital of the jth energy level of the adsorbate molecule; ε_F is the Fermi

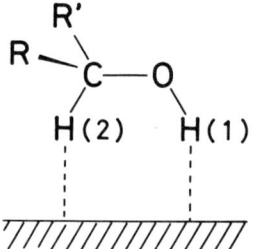

Fig. 30 A model of an alcohol molecule adsorbed on the surface of a liquid metal [2,46].

level of the liquid metal; ε_j is the jth energy level of the adsorbate molecule; ε_k is the kth energy level of the liquid metal; X_s is the atomic orbital of the sth hydrogen atom in the adsorbate; φ_k is the wave function of the liquid metal corresponding to ε_k; and H' is the perturbation potential representing the interaction between the catalyst surface and adsorbate molecule. In addition, the summations Σ_j^{unocc} and Σ_j^{occ} mean summations over unoccupied levels and occupied levels of the adsorbate, respectively, and $\Sigma_{k,\varepsilon_k=\varepsilon_F}$ indicates a summation over the levels near the Fermi level of the liquid metal.

The first term and the second term in the bracket of the right-hand side of Eq. 33 are abbreviated by D_N and D_E, respectively. It is convenient to define $D_{N,H(1)}$ and $D_{N,H(2)}$, which denote the delocalization of electron from metal to the adsorbing molecule through H(1) and the electron delocalization through H(2) atom, respectively. Similarly, the electron delocalizations through the reverse routes are expressed by $D_{E,H(1)}$ and $D_{E,H(2)}$.

Numerical values of $D_{N,H(1)}$, $D_{N,H(2)}$, $D_{E,H(1)}$, and $D_{E,H(2)}$ are shown in Fig. 31 as a function of Fermi energy. It is clear that the electron delocalization from metal to adsorbate through H(1) atom predominates over the delocalization through H(2) atom. In addition, it is seen that the electron delocalization from

Electronic Aspect of Catalysis

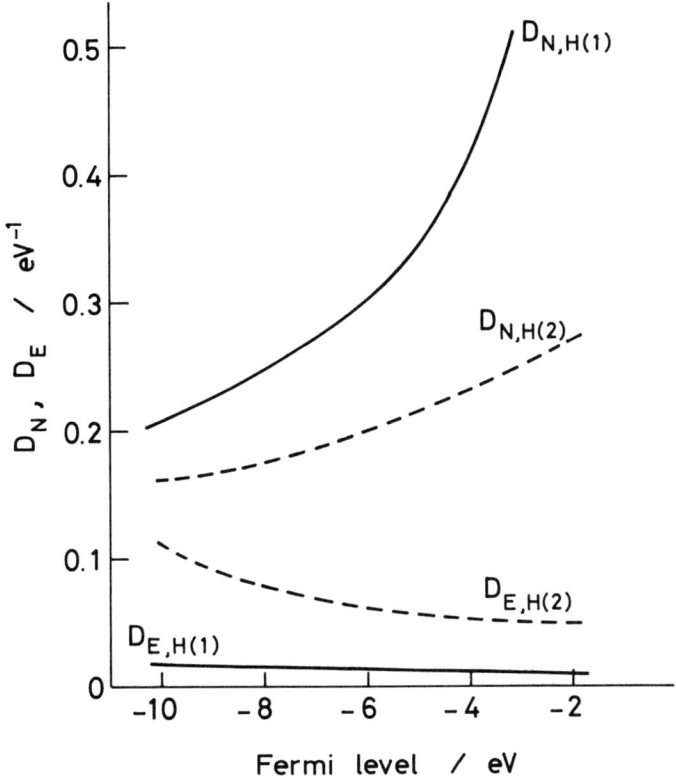

Fig. 31 Magnitudes of the electron delocalization parameters D_N and D_E as a function of the position of the Fermi level [2,46].

metal to adsorbate becomes easier as the Fermi level ε_F of the metal is elevated. Of course, the reverse is true for the delocalization of electrons from adsorbate to the catalyst metal, as represented by the depression of D_E with increasing ε_F. It must be pointed out that the delocalization of electrons from adsorbate to metal through the H(2) atom predominates over that through the H(1) atom.

The above discussion explains the fact that the metals with smaller work function are catalytically active. From the energy

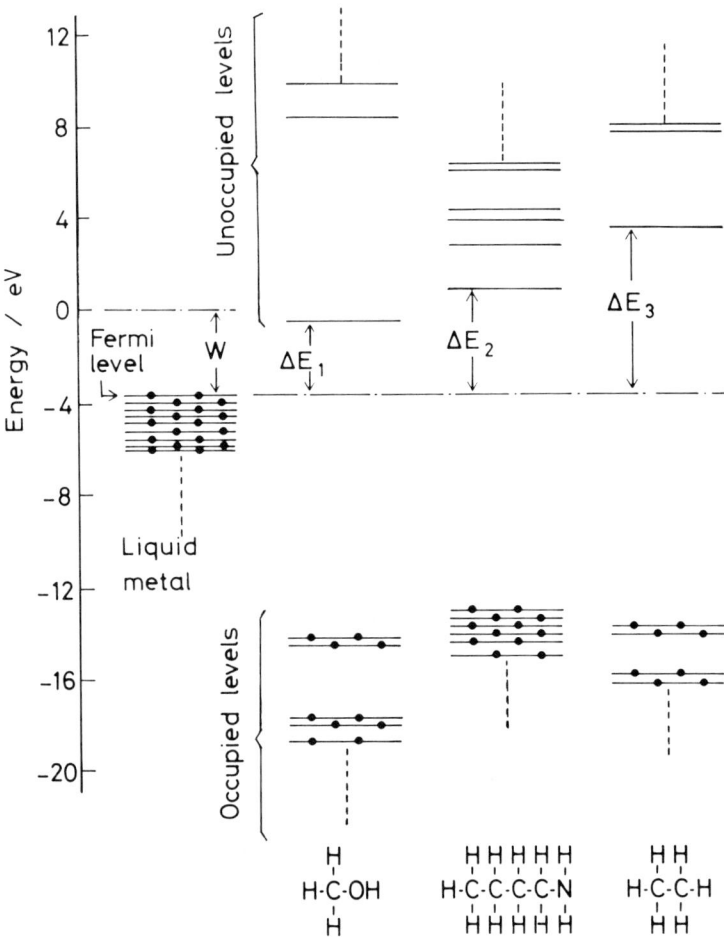

Fig. 32 Electronic energy levels for methanol, *n*-butylamine, ethane, and a liquid metal [2,11,46].

diagram shown in Fig. 32 [2,46], it can be easily understood that the smaller the work function the larger the extent of the electron delocalization from the highest occupied level of the catalyst metal to the lowest unoccupied level of the reactant alcohol.

Finally, it must be added that almost the same discussion as above is possible for the dehydrogenation of amines over the liquid metals [2,46].

2. Electronic Aspects of the Catalysis of Liquid Tellurium

The aforementioned discussion can not be applied to the catalysis by Te(liq.). It is hard to consider that the electron-donating ability of Te(liq.) is large because tellurium has a large first ionization potential. In connection with this, it must be pointed out that, strictly speaking, tellurium is a nonmetal. Thus, it is not unlikely that the character of the catalysis of Te(liq.) is different from that of the metallic liquid. Indeed, Te(liq.) is active only for the dehydrogenation of polynuclear hydrocarbons (Chapt. 3, IV) [11,53]. Neither alcohols nor hydrocarbons other than polynuclear hydrocarbons can be dehydrogenated over the Te(liq.) catalyst.

In the light of theoretical energy calculations based on the model shown in Fig. 33a-d, a sharp contrast appears between the catalytic activity of Te(liq.) and that of the metallic liquid [11,44]. In each of these models, the model catalyst consists of two atoms; the interatomic distance between these two atoms is equal to the nearest neighbor distance obtained from the radial distribution function of the catalyst liquid. By applying a quantum chemical formalism (the extended Hückel method), it is possible to calculate the electronic interaction energy as a function of the distance $\bar{r}$ between the model catalyst and the hydrogen atoms to be removed from the reactant molecule.

The results of calculation are shown in Fig. 34a'-d' [11, 44]. It can be seen that the energy curve for the tellurium-tetralin system (Fig. 34a') has a minimum whereas the curve for the indium-tetralin system (Fig. 34b') has no minimum. On the contrary, the curve for the tellurium-methanol system (Fig. 34c') has no minimum, and the curve for the indium-methanol system

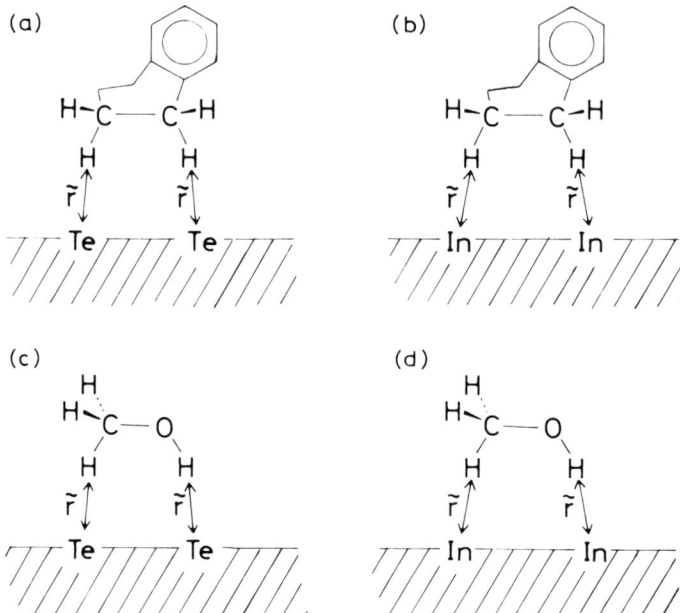

Fig. 33 Catalysis models assumed for calculating the interaction energies: (a) tetralin-Te(liq.), (b) tetralin-In(liq.), (c) methanol-Te(liq.), (d) methanol-In(liq.) [11,44].

(Fig. 34d') has a minimum. It appears reasonable to assume that when the energy curve for the model catalyst exhibits a minimum, the corresponding real catalyst is capable of adsorbing and activating the reactant molecule. We can therefore expect that Te(liq.) is active for the tetralin dehydrogenation but it is not active for the methanol dehydrogenation. Similarly, we can expect that In(liq.) is active for the methanol dehydrogenation but is not active for the tetralin dehydrogenation. All these expectations are consistent with the experimental results.

The theoretical treatment mentioned above has brought about additional information: The nearer approach of the reactant molecule to the model tellurium catalyst results in an increase

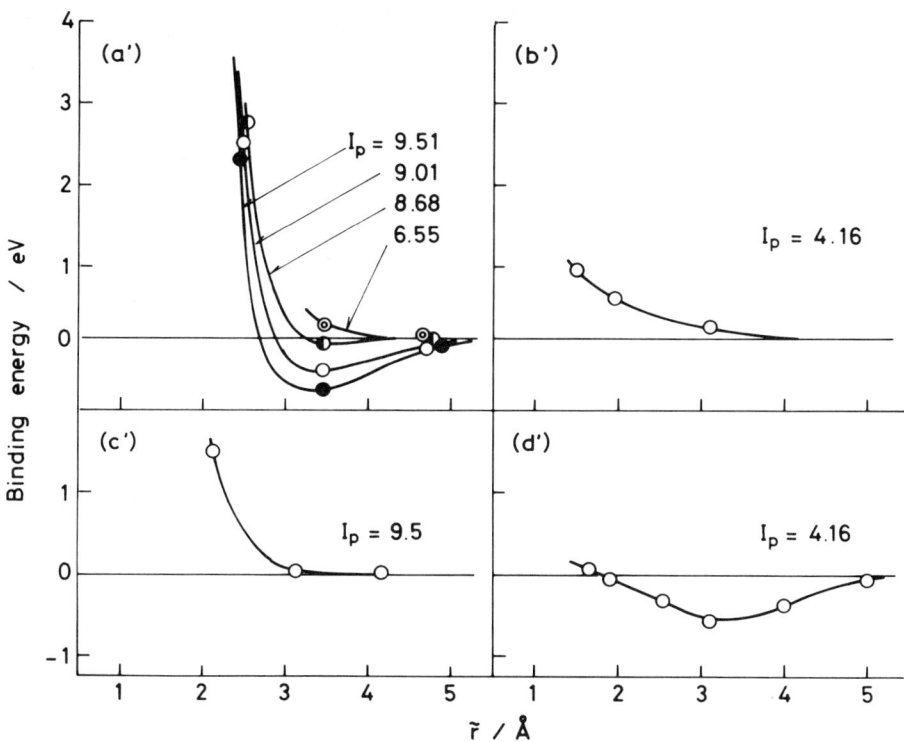

Fig. 34 The catalyst-reactant interaction energies (I_p is the first ionization potential (eV) of the catalyst): (a') tetralin-Te (liq.), (b') tetralin-In(liq.), (c') methanol-Te(liq.), (d') methanol-In(liq.) [11].

in the density of the negative charge on the tellurium atom [11, 44]. This is just the reverse of what we can expect for the electron transfer involved in the reaction over the metallic liquid.

Presumably the high ionization potential of tellurium is favorable for the electron transfer from tetralin to tellurium, thus causing the stabilization of the system. This result is very important. It is more a prediction than a verification. It provides us with the way to improve the catalytic activity of Te(liq.): If we can enhance the apparent ionization potential of Te(liq.), we can get a good catalyst. Since the ionization potential of selenium is higher than that of tellurium and, in addition, these two elements are capable of forming a solid solution with a low melting point at any composition, the Te-Se liquid mixture is a good example for examining whether we can make a good catalyst by the way mentioned above. According to the experimental results (Table 3, Chapt. 3, IV) [11,54], the incorporation of selenium into Te(liq.) brings about a high catalytic activity for the dehydrogenation of polynuclear hydrocarbons.

REFERENCES

1. A. Miyamoto and Y. Ogino, *J. Catal.*, *27*, 311 (1972).
2. A. Miyamoto, Ph.D. Thesis, Tohoku University, 1975.
3. Y. Saito, F. Miyashita, and Y. Ogino, *J. Catal.*, *36*, 67 (1975).
4. Y. Saito, Ph.D. Thesis, Tohoku University, 1978.
5. S. Glasstone, K. J. Laidler, and H. Eyring, *The Theory of Rate Processes*, McGraw-Hill Book Company, Inc., New York and London, 1941, p. 351.
6. C. Kemball, *Proc. Roy. Soc.*, *A190*, 117 (1947).
7. A. Miyamoto and Y. Ogino, *J. Catal.*, *37*, 133 (1975).
8. A. Miyamoto and Y. Ogino, *J. Catal.*, *41*, 212 (1976).
9. P. J. Robinson and K. A. Holbrook, *Unimolecular Reactions*, Wiley-Interscience, London, New York, Sydney, Toronto, 1972, p. 287.
10. W. A. Van Hook, in *Isotope Effects in Chemical Kinetics* (C. J. Collins and N. S. Bowman, eds.), Van Nostrand Reinhold Company, New York, Cincinnati, Toronto, London, Melbourne, 1970, p. 1.
11. K. Takahashi, Ph.D. Thesis, Tohoku University, 1979.
12. K. Takahashi and Y. Ogino, *Fuel*, *60*, 975 (1981).
13. C. T. Ewing, J. P. Stone, J. R. Spann, and R. R. Miller, *J. Phys. Chem.*, *71*, 473 (1967).
14. S. Ozawa, T. Suenaga, and Y. Ogino, *Fuel*, *64*, 712 (1985).

References

15. S. Ozawa, K. Sasaki, and Y. Ogino, *Fuel*, *65*, 707 (1986).
16. D. C. Cronauer, Y. T. Shah, and R. G. Ruberto, *Ind. Eng. Chem. Process. Des. Dev.*, *17*, 281 (1978).
17. S. Ozawa, M. Matsuura, S. Matsunaga, and Y. Ogino, *Fuel*, *63*, 719 (1984).
18. D. Marquardt, *J. Soc. Ind. Appl. Math.*, *11*, 431 (1963).
19. Y. Ogino, S. Ozawa, and K. Ishikawa, to be published in *Fuel Processing Technology* (Y. Sanada, ed.), Elsevier Scientific Publishing Company, Amsterdam.
20. A. Miyamoto and Y. Ogino, *J. Catal.*, *43*, 143 (1976).
21. H. Sugawara, Ms. Thesis, Tohoku University, 1980.
22. H. Sugawara and Y. Ogino, *J. Chem. Soc. Faraday Trans. 1*, *78*, 1079 (1982).
23. Y. Saito and Y. Ogino, *J. Catal.*, *55*, 198 (1978).
24. L. A. Popova, S. I. Popel, and Y. I. Maslennikov, *Russ. J. Phys. Chem.*, *55*, 993 (1981).
25. L. A. Zhukova and S. I. Popel, *Russ. J. Phys. Chem.*, *56*, 1661 (1982).
26. M. L. Josi, *Rev. Sci. Instrum.*, *36*, 678 (1965).
27. B. R. Orton, B. A. Shaw, and G. I. Williams, *Acta Metall.*, *8*, 177 (1960).
28. R. Kaplow, S. L. Strong, and B. L. Averbach, *Phys. Rev.*, *138*, A 1336 (1965).
29. J. E. Enderby, D. M. North, and P. Egelstaff, *Phil. Mag.*, *14*, 961 (1966).
30. Y. Waseda and M. Ohtani, *Sci. Rep. Res. Inst. Tohoku University*, *23*, 188 (1972).
31. D. T. Cromer, *Acta Cryst.*, *18*, 17 (1965).
32. C. N. J. Wagner and N. C. Halder, *Adv. Phys.*, *16*, 241 (1967).
33. C. N. J. Wagner, in *Liquid Metals Chemistry and Physics* (S. Z. Beer, ed.), Marcel Dekker, Inc., New York, 1972, p. 257.
34. S. P. Isherwood and B. R. Orton, *J. Appl. Cryst.*, *2*, 219 (1969).
35. V. K. Semenchenko, *Surface Phenomena in Metals and Alloys*, Pergamon, New York, 1961, pp. 43-56.
36. Y. Saito, H. Yoshida, T. Yokoyama, and Y. Ogino, *J. Colloid Interf. Sci.*, *66*, 440 (1978).

37. R. Hultgren, *Selected Values of Thermodynamic Properties of Metals and Alloys*, Wiley, New York, 1963.
38. Y. Saito, H. Yoshida, A. Miyamoto, T. Yokoyama, and Y. Ogino, *J. Catal.*, *55*, 36 (1978).
39. V. F. Kovalchuk, *Zhur. Fiz. Khim.*, *43*, 184 (1969).
40. V. G. Ryabov, *Izv. Vyssh. Uchbeb Zaved. Isvet. Met.*, *14(2)*, 82 (1971).
41. T. P. Hoar, *Trans. Faraday Soc.*, *53*, 315 (1957).
42. H. Tsukamoto, Ms. Thesis, Tohoku University, 1985.
43. M. Balooch, W. J. Siekhaus, and D. R. Olander, *J. Phys. Chem.*, *88*, 3521 (1984).
44. Y. Ogino, *Catal. Rev.-Sci. Eng.*, *23*, 505 (1981).
45. C. C. Addison, *The Chemistry of Liquid Alkali Metals*, John Wiley and Sons, Chichester, New York, Brisbane, Toronto, Singapore, 1984, pp. 193-211.
46. A. Miyamoto, K. Okano, and Y. Ogino, *J. Catal.*, *36*, 276 (1975).
47. C. Norris, *Inst. Phys. Conf. Ser.*, *30*, 171 (1977).
48. K. Honda and Y. Ogino, *J. Chem. Soc. Chem. Commun.*, 332 (1980).
49. K. B. Khokonov, S. N. Zadumkin, and B. B. Alchagirov, *Sov. Electrochem.*, *10*, 865 (1974).
50. T. E. Faber, *An Introduction to the Theory of Liquid Metals*, Cambridge University Press, 1972, p. 481.
51. T. Masuda, Ms. Thesis, Tohoku University, 1980.
52. S. Ohsaka, Ms. Thesis, Tohoku University, 1982.
53. K. Takahashi and Y. Ogino, *Chem. Lett.*, 423 (1978).
54. K. Takahashi and Y. Ogino, *Chem. Lett.*, 549 (1978).

5
Surface Tension of Liquid Metals and Alloys

I.	Introduction	125
II.	Various Problems Related to Surface Tension	126
	A. Progress in the Theory	126
	B. Progress in Experimental Techniques	127
	C. Information from Surface Tension	131
	D. Multicomponent System	134
	E. Correlation of Surface Tension with Other Properties	148
	References	150

I. INTRODUCTION

As we have seen in the preceding chapter, we need information about a number of subjects, in particular about the surface properties of liquid metals and liquid alloys, in order to have a deep understanding on the surface catalysis. Unfortunately, only pieces of information about surface properties of liquid metals are available in books already published [1-3]. Thus, it is pertinent to review and integrate the principal research done thus far regarding the structures and properties of liquid metals and liquid alloys.

For this purpose, this chapter reviews surface tension studies. Surface tension is an important classical physical quantity that can be applied to adsorption study and to the determination of surface composition of liquid alloys, as we have already seen in the preceding chapter (Chapt. 4, IV, C, 1). Because Allen [4] has already reviewed the same subject and provided us with information accumulated before 1970, the focus of this review centers on the progress made in the years since.

II. VARIOUS PROBLEMS RELATED TO SURFACE TENSION

A. Progress in the Theory

Surface tension is an important physical quantity of a substance in the liquid state, and hence intensive theoretical studies [5] have been made for a long time. Excellent reviews on the theory of surface tension of ordinary liquids have been published by Ono and Kondo [6], Croxton and Ferrier [7,8], Abraham [9], and Evans [10]. The theories for metallic liquids have also been reviewed by Semenchenko [11], Faber [1], Shimoji [3], and by Croxton [12]. Thus, we have ample sources of information about the theory of surface tension of liquid metals and liquid alloys. Therefore only important progress achieved in the recent years is reviewed below.

In interpreting experimental results, several researchers have employed the classical surface tension equation proposed by Fowler [13]

$$\gamma = \frac{\pi n^2}{8} \int_0^\infty r^4 \frac{\partial u(r)}{\partial r} g_0^{(2)}(r)\,dr \tag{1}$$

where n is the number density of atoms of a metallic liquid, r is the distance from any specified central atom, u(r) is the pair interaction potential, and $g_0^{(2)}(r)$ is the pair correlation function. Waseda and Ohtani [14] have determined u(r) and $g_0^{(2)}(r)$ from the X-ray scattering data and have evaluated the surface tensions of Cu(liq.), Ag(liq.), Au(liq.), Fe(liq.), Co(liq.), and

Ni(liq.). By the same procedures, Belan-Gayko et al [15] have determined the surface tension of Al(liq.). Lepinskikh and Stepanova [16] have somewhat improved Fowler's equation in order to evaluate the surface tensions of various liquid metals and liquid alloys. Unfortunately, the above mentioned research has been unsuccessful in obtaining proper values of the surface tension. This is due mainly to a deficiency of Fowler's equation wherein special roles of electrons in exhibiting the surface tension are ignored.

Advanced theories of the surface tension of liquid metals are developed under the recognition of the special contribution of electrons to the surface tension. Many theoretical papers have been published [17-30] since Evans's work of 1974 [31]. Among these theories, that of Hasegawa-Watabe [22] deserves attention because it enables us to calculate a surface tension that agrees well with the experimental results: The theoretical values agree with the experimental values for Li(liq.), Na(liq.), K(liq.), Rb(liq.), Cs(liq.), Mg(liq.), Zn(liq.), Cd(liq.), Al(liq.), and Pb(liq.). Similarly, Chacon et al [32] have reported that the theory proposed by them agrees with the experimental results for Li(liq.), Na(liq.), K(liq.), Cs(liq.), Mg(liq.), Zn(liq.), and Al(liq.).

B. Progress in Experimental Techniques

Earlier experimental data reported before 1970 have been summarized by Allen [4]; the data reported from 1970 to 1985 have been collected and part of them are presented in Table 1 [33-44].

As we can see in Table 1, most of the experimental data have been obtained either by the maximum bubble pressure method or by the sessile drop method. Principles of these two methods, together with those of other methods, are described in the book published by Semenchenko [11], and more particular experimental precautions are given by White [45]. It has been pointed out by White that the thermodynamic equilibrium is an important prerequisite for obtaining surface tension data that deserve

TABLE 1 Surface Tension Data for Various Liquid Metals

Liquid metals	γ_m (dyn/cm) [a]	Accuracies (dyn/cm)	$-d\gamma/dT$ (dyn/cm deg)	Temperature (°C)	Methods	Refs.
Cu	1555[b]	max ± 20	0.233	1083-1450	SD (Ar; H_2)	33
	1390 ± 32	—	0.43	1000-1330	OD (H_2)	34
	1552[c] ± 35	—	0.176 ± 0.023	1100-1588	SD (Ar; 7.5×10^{-8} Torr)	35
	1290	±7	0.16	1090-1270	MBP (Ar)	36
	1282	±5	0.15	1093-1364	SD (N_2 + 5% H_2)	36
Ag	1142[b]	max ± 20	0.185	961-1450	SD (Ar; H_2)	33
Au	1452[b]	max ± 20	0.251	1063-1450	SD (Ar; H_2)	33
Zn	806	—	0.25	420-800	SD (10^{-9} Torr base pressure)	37
	868	±3	0.15	690-915	MBP (2×10^{-6} Torr)	36
Al	865	±6	0.15	694-904	SD (2×10^{-6} Torr)	36
	865	±4	0.12	700-1000	SD (Ar)	38
	714	—	0.088	—	SD (10^{-9}-10^{-10} Torr)	39
Ga	(γ = 725, 723.7, 723.9, 724.5, and 723.2 dyn/cm at 17, 25, 29.6, 30, and 40°C, respectively)				PD	40
	556	—	0.081	—	SD (10^{-9}-10^{-10} Torr)	39
In	553	±2	0.13	174-450	MBP (2×10^{-6} Torr)	36
	558	±3	0.11	308-586	SD (2×10^{-6} Torr)	36

	γ_m[a]		Temp. range	Method (Atmosphere)	Ref.
	500[b]	–	327-1100	SD (Ar; H_2)	33
Pb	444.2 ± 1.0	0.094 ± 0.003	327.5-877	SD (10^{-9} Torr base pressure)	42
	470.0 ± 2.0	0.164	mp.-540	MBP (Ar)	41
	581[b]	–	232-1450	SD (Ar; H_2)	33
Sn	552 ± 2.0	0.167	mp.-500	MBP (Ar)	41
	551 ±7	0.17	294-546	MBP (Ar)	36
	561 ±7	0.13	298-702	SD (Ar)	36
	380 ± 1.0	0.142	mp.-560	MBP (Ar)	41
Bi	376 ±4	0.070	346-565	MBP (Ar)	36
	372 ±2	0.089	286-719	SD (Ar)	36
	370	0.090	–	SD (10^{-9}-10^{-10} Torr)	39
Fe	1918 ± 30	0.43 ± 0.06	1508-1832	OD (H_2; H_2 + He)	43
	1729 + 1.76(t - t_m) - 2.21 × 10^{-4}(t^2 - t_m^2);		–	PD (Ar)	44

(t, temperature in °C; t_m, melting point (mp.) in °C)

[a] γ_m: Surface tension at the melting point T_m (K) or t_m (°C).
[b] Value of γ_0 defined by $\gamma = \gamma_0 + (d\gamma/dt)t$; t in °C.
[c] Value of γ_0 defined by $\gamma = \gamma_0 + (d\gamma/dT)T$; T in K.
[d] SD, sessile drop; OD, oscillating drop; MBP, maximum bubble pressure; PD, pendant drop. Atmospheres are shown in parentheses.

theoretical analysis. Although this point is still unclear in the reports listed in the table, many improvements in the experimental techniques can be seen in the reports [46,47]. Compared with the older work, recent experimental studies of surface tension pay particular attention to preventing surface contaminations: The use of extremely pure working gas or measurements under high vacuum conditions is reported.

An application of the levitation technique to the surface tension measurement by the oscillating drop method also aims at preventing surface contaminations, in particular from the contamination due to the contact of the sample liquid with the specimen holder [34]. In this method a drop of the sample liquid is suspended in a levitation cell by an external electromagnetic force and oscillatory changes in the shape of the sample drop are detected optically and amplified. The surface tension γ is evaluated by the following Rayleigh's equation

$$\gamma = \frac{3}{8} \pi m \omega^2 \tag{2}$$

where m is the mass of the drop and ω is the frequency of the oscillation.

It is well known that the most serious contaminations are the incorporations of sulfur (S) and oxygen (O) into the sample [45]. Small amounts of oxygen incorporation usually result in a great depression of the surface tension, e.g., Ga(liq.) [40], Cu(liq.) [35], and Al(liq.) [48]. Little surface tension change due to the oxygen contamination for Pb(liq.) [42] appears exceptional. In the case of Cu(liq.), it is reported that the lower the temperature the larger the surface tension depression due to the oxygen contamination [49]. In addition, the sign of the temperature dependence of the surface tension of Cu(liq.) changes from negative to positive when the surface contamination due to oxygen increases [49]. This fact is important in discussing the surface entropy of the liquid metal.

From the reason mentioned above, we have to pay special attention to the source of oxygen contaminants. It is reported that very small amounts of oxygen can enter the surface tension cell by diffusion through the nonmetallic part of the apparatus [48]. Chemical reactions between the working gas and impurities contained in the specimen holder can also provide the sample with oxygen;

$$SiO_2(\text{solid}) + H_2(\text{gas}) \longrightarrow SiO(\text{gas}) + H_2O(\text{gas}) \quad (3)$$

$$SiO(\text{gas}) \longrightarrow Si + O \quad (4)$$

C. Information from Surface Tension

1. Width of Transition Zone

Although the surface tension data involve information about the structure of the surface of the liquid metal, current theories are still incomplete and mostly unavailable for analyzing the surface structure except for the surface composition. Nevertheless, a few researchers have examined extracting structural information for the liquid metal surface from surface tension data. For instance, Brown and March [50] have derived the following relation

$$\chi_T \gamma = \frac{3}{4} \ell \quad (5)$$

where χ_T is the isothermal compressibility of the sample liquid, and ℓ is the width of the transition zone that is postulated to exist between the metallic phase and the vapor phase in contact with the liquid metal (Chapt. 6). Similarly, using the radial distribution function obtained by the X-ray scattering experiment (Chapt. 4, IV, B), Wasewda and Jacob [51] derived the following relation

$$\gamma \chi_T = (0.058 \pm 0.007)\ell \quad (6)$$

This equation gives a reasonable value for the width of the transition zone, as we can see in Table 2.

TABLE 2 Width of Transition Zone and Surface Entropy for Various Liquefied Elements

Element	Transition[a] zone (Å)	Surface entropy (dyn/cm·deg)			
		[b]	0.16[c]	0.11[d]	0.14[e]
Li	5.7	-	0.16	0.11	0.14
Na	5.8	0.19	0.14	0.08	0.06
K	6.9	0.12	0.07	0.05	0.03
Rb	7.0	0.11	0.06	0.04	-
Cs	8.2	0.09	0.05	0.04	-
Cu	3.1	0.40	0.28	0.17	0.30
Ag	3.2	0.31	0.24	0.15	0.22
Au	3.0	-	0.27	0.13	0.24
Mg	4.8	-	0.24	0.19	0.12
Ca	6.3	-	0.14	0.10	0.03
Ba	7.0	-	0.06	0.08	0.04
Zn	3.3	0.34	0.25	0.28	0.14
Cd	3.4	0.27	0.23	0.24	0.11
Hg	3.2	0.26	0.22	0.34	0.13
Al	3.4	0.31	0.26	0.11	0.38
Ga	3.0	0.31	0.18	0.10	0.13
In	3.3	0.25	0.17	0.09	0.02
Tl	3.3	0.24	0.13	0.11	0.20
Ge	-	-	0.14	0.07	0.05
Sn	3.3	0.24	0.14	0.07	0.13
Pb	3.4	0.22	0.13	0.10	0.11
Sb	3.4	0.23	0.09	0.09	0.02
Bi	3.4	0.21	0.11	0.08	0.02
Fe	3.3	0.40	0.47	0.21	0.44
Co	2.9	0.41	0.40	0.20	0.39
Ni	2.9	0.41	0.45	0.19	0.35

[a] Ref. 51.   [c] Ref. 33.   [e] Ref. 56.
[b] Ref. 52   [d] Ref. 55.

2. Surface Entropy

The temperature dependence of the surface tension of a liquid metal is expected to provide us with valuable information because it is related to the excess surface entropy $\hat{S}$ per unit of the surface area [12,45]

$$\frac{d\gamma}{dT} = -\hat{S} \tag{7}$$

If we have a liquid metal with $d\gamma/dT > 0$, it is inferred that the surface is more ordered than the bulk of the liquid metal. Such an anomalous surface structure is of special interest if it really exists. However, we cannot find such a metal as shown in Table 2. In connection with this, we have to remember that the temperature dependence of surface tension is very sensitive to the surface contamination. The sign of $d\gamma/dT$ for Cu(liq.) is reported to change from negative to positive when the extent of the surface contamination due to oxygen is increased [49].

The theoretical work on the temperature dependence of the surface tension is also worth mentioning. Itami and Shimoji [52] have treated this problem within the framework of Evan's theory [31]. By assuming a rigid sphere model for the ionic distribution in the liquid metal, they approximated γ by the sum of an electronic term γ_e and an ionic term γ_H

$$\gamma = \gamma_e + \gamma_H \tag{8}$$

Since the temperature dependence of γ_e is expected to be small, it is possible to make an approximation that $\partial\gamma/\partial T \doteqdot \partial\gamma_H/\partial T$. Thus, we can use Fowler's equation as an explicit expression for γ_H. Furthermore, it is possible to use the following equation of states [53]

$$g_0^{(2)} = \frac{(2 - \xi)}{2(1 - \xi)^3} \tag{9}$$

where $\xi = \pi\sigma^3 n/6$ and σ is the hard sphere diameter. The final expression obtained by Itami for the surface entropy $\hat{S}$ is thus

$$\hat{S} = \frac{\pi n^2 k(2-\xi)\sigma^4}{16(1-\xi)^3}\left[1 + \frac{4T}{\sigma}\left(\frac{\partial\sigma}{\partial T}\right)_v\right]$$

$$+ \frac{3\pi n^2 kT(5-2\xi)\sigma^3 \xi}{16(1-\xi)^4}\left(\frac{\partial\sigma}{\partial T}\right)_v \tag{10}$$

where k is Boltzmann's constant and the subscript v is the volume.

As we can see in Table 2, the values of $\hat{S}$ calculated by Eq. 10 with an assumption that $(\partial\sigma/\partial T)_v = 0$ are reasonable in magnitudes. A report similar to that mentioned above has been published by Tewari et al [54].

D. Multicomponent System

1. *Surface Tension of Liquid Alloys*

A number of reports on the surface tension of binary liquid alloys have been published since 1970, as we can see in Table 3. Experimental techniques and precautions for measurements are similar to those mentioned for the single component system. It can be seen in the table that the maximum bubble pressure method (MBP) and the sessile drop method (SD) are adopted almost exclusively. Besides the data listed in the table, the results of surface tension measurements for the following binary liquid alloys have been reported: SD (in Ar) for Al-Bi, Zn-Bi, Zn-Pb [38]; SD (in vacuum of 10^{-5} Torr, 232-700°C) for Pb-Sn (30-100 wt% Sn) [60]; MBP or CD (capillary depression method) for Zn-Bi (723-903 K) [61]; SD or MBP (up to 1923 K) for Fe-As (3.7-19.5 atom% As) [62]; large-drop method (in He) for Pt-Sn, Pt-Al, Pt-Si, Pt-Pd, and Pt-Co [63].

Surface tension data for ternary liquid alloys are also seen in literature. For instance, Savvin and Ibragimov [64] have measured the surface tension of the Bi-Pb-Hg liquid alloy using the MBP method and summarized the data into the following three formulae:

Problems Related to Surface Tension

$$446.8 + 76561 \ln(1 + 0.06x_{Hg}) - 4491x_{Hg} \quad (300°C)$$
$$431.3 + 4046 \ln(1 + 0.3x_{Hg}) - 1103x_{Hg} \quad (350°C) \quad (11)$$
$$427.9 + 33288 \ln(1 + 0.1x_{Hg}) - 3231x_{Hg} \quad (400°C)$$

where x_{Hg} is the mole fraction of mercury: The ratio of mole fraction of Bi to that of Pb is kept constant, i.e., $x_{Bi}/x_{Pb} = 5/95$.

Other ternary systems served for the surface tension studies are as follows: Ag-Au-Cu (SD in Ar or Ar + H_2) [65], Ga-In-Sn (MBP) [66], and Fe-Ni base alloy + X (X = Sb, Se, or Te) (MBP in Ar) [67].

2. Surface Segregation

In the discussion of the surface tension of multicomponent liquid alloys, we have to take the surface segregation into consideration. The surface segregation frequently brings about a large difference between the bulk composition and surface composition. As we have already seen in the preceding chapter (Chapt. 4, IV, C, 1), the surface composition is related to the surface catalysis. Thus, the theoretical basis of studying the surface segregation is given below in detail.

Now let us consider, for convenience, a binary liquid alloy A-B (A and B are the two components of the alloy). Then we can obtain the following relation at a constant temperature

$$d\gamma = -\Gamma_A d\mu_A^b - \Gamma_B d\mu_B^b \quad (12)$$

where Γ_j is the molar number of atoms (or molecules) of the j-component existing in unit area of the surface, and μ_j^b is the chemical potential of the j-component in the bulk of the liquid alloy.

Combining Eq. 12 with the Gibbs-Duhem relation and rearranging, we obtain

TABLE 3 Surface Tension Data for Several Binary Liquid Alloys

Liquid alloys A-B	Composition (A-atom%)	$\gamma^{a,b}$ (dyn/cm)	Accuracies (dyn/cm)	Temperature (°C)	Methods	Ref.
Pb-Sn	0	$626 - 0.222t$		608-800	MBP (Ar)	57
	20	$525 - 0.127t$		600-800		
	26.1	$471 - 0.085t$	$\pm 1^c$	600-887		
	50	$520 - 0.182t$		600-800		
	70	$552 - 0.235t$		603-800		
	100	$548 - 0.2335$		650-800		
Pb-Sn	0	$566.84 - 4.76 \times 10^{-2} t$	± 1.9	240-844	MBP (He)	58
	26.8	$516.3 - 5.3 \times 10^{-2} t$	± 7.0	195-570		
	39.6	$498.1 - 4.5 \times 10^{-2} t$	± 1.9	211-568		
	56.6	$489.2 - 6.6 \times 10^{-2} t$	± 3.0	269-530		
	70.0	$478.4 - 6.3 \times 10^{-2} t$	± 5.0	297-498		
	100	$467.7 - 6.6 \times 10^{-2} t$	± 2.1	344-652		
Al-Pb	100	$865 - 0.12(t - t_m)$	± 4		SD	38
	99.9935	$908.8 - 0.089(t + 273)$	± 6			
	99.93	$103.8 + 0.584(t + 273)$	± 7			
	99.89	$421.8 + 0.25(t + 273)$	± 7	700-1000		

Problems Related to Surface Tension

System	Composition	Equation	Error	Temp. range	Method	Ref.
	99.84	$354.3 + 0.26(t + 273)$	±4			
	99.7	$46.99 + 0.49(t + 273)$	±6			
	99.6	$722.35 - 0.12(t + 273)$	±7			
	0	$462 - 0.11(t - t_m)$	±6			
	100	$865 - 0.12(t - t_m)$	±4			
	99.954	$853.6 - 4.17 \times 10^{-2}(t + 273)$	±7			
Al-Sn	99.54	$793.2 + 4.46 \times 10^{-3}(t + 273)$	±9	700-1050	SD (10^{-6} Torr)	38
	98.86	$526.4 + 0.191(t + 273)$	±9			
	97.53	$541.2 + 0.134(t + 273)$	±7			
	96.2	$640.9 - 1.6 \times 10^{-2}(t + 273)$	±6			
	0	$561 - 0.13(t - t_m)$	±7			
	100	$868 - 0.15(t - t_m)$	±3			
	97.4	$984 - 0.18t$	±3			
	90.3	$997 - 0.14t$	±2			
	84.3	$948 - 0.11t$	±2			
	82.3	$956 - 0.11t$	±3			
Al-Cu	77.9	$947 - 0.10t$	±3	1100	MBP (Ar)	59
	70.3	$950 - 0.09t$	±3			
	61.0	$984 - 0.08t$	±5			
	51.1	—	—			
	44.2	1029	—			
	37.0	1122	—			
	29.6	1190	—			

TABLE 3 (continued)

Liquid alloys A-B	Composition (A-atom%)	$\gamma^{a,b}$ (dyn/cm)	Accuracies (dyn/cm)	Temperature (°C)	Methods	Ref.
	23.2	—				
	18.1	1230	—			
Al-Cu	7.9	1230	—	1100	MBP (Ar)	59
	3.5	—	—			
	0	$1290 - 0.16(t - t_m)$	—			
	100	$865 - 0.15(t - t_m)$	±6			
	97.3	$1012 - 0.18t$	±4			
	90.6	$981 - 0.14t$	±5			
	84.0	$1035 - 0.18t$	±9			
	82.4	—	—			

Problems Related to Surface Tension

Al-Cu			
78.7	1084 − 0.22t	±8	Different temperatures for different compositions
70.1	1098 − 0.17t	±7	
60.3	1107 − 0.15t	±6	SD
49.9	1094 − 0.08t	±4	
43.5	1096 − 0.04t	±4	
36.6	908 + 0.16t	±4	
28.1	1025 + 0.15t	±4	
21.1	—	—	
7.0	1346 − 0.13t	±8	
0	1282 − 0.15(t − t_m)	±5	59

^{a}t in °C.
^{b}t_m, melting point in °C.
^{c}Relative error in % (only this value; others are standard deviation in dyn/cm).

$$-\left(\Gamma_B - \Gamma_A \frac{x_B^b}{x_A^b}\right) = \frac{1}{RT}\left(\frac{\partial \gamma}{\partial \ln a_B^b}\right)_T \quad (13)$$

where x_j^b is the atomic fraction of the jth component in the bulk of the alloy and a_j^b is the bulk activity of the jth component.

If we define a new quantity $\tilde{\Gamma}_B$ by

$$\tilde{\Gamma}_B = \Gamma_B - \Gamma_A \frac{x_B^b}{x_A^b} \quad (14)$$

then the Gibbs adsorption isotherm results, i.e.,

$$-\tilde{\Gamma}_B = \frac{1}{RT}\left(\frac{\partial \gamma}{\partial \ln a_B^b}\right)_T \quad (15)$$

A multiplication of x_A^b to $\tilde{\Gamma}_B$ again defines a new quantity $\hat{\Gamma}_B$, which is related to the surface composition by

$$\hat{\Gamma}_B \equiv x_A^b \tilde{\Gamma}_B = \frac{-x_A^b}{RT}\left(\frac{\partial \gamma}{\partial \ln a_B^b}\right)_T = \left(\frac{N_s}{A}\right)(x_B^s - x_B^b) \quad (16)$$

where N_s is the sum of the molar number of the surface A atoms ($N_{A,s}$) and that of the surface B atoms ($N_{B,s}$), A is the surface area, and x_B^s is the atomic fraction of B atoms in the surface, i.e.,

$$x_B^s = \Gamma_B/(\Gamma_A + \Gamma_B) = N_{B,s}/(N_{A,s} + N_{B,s})$$

It must be pointed out that a simple relation exists between $\hat{\Gamma}_A$ and $\hat{\Gamma}_B$

$$\hat{\Gamma}_A = -\hat{\Gamma}_B \quad (17)$$

The equations mentioned above are useful for calculating the surface composition from the surface tension data. For instance, Semenchenko [11] has utilized Eq. 16 to derive an expression for the surface atomic fraction of B, i.e.,

$$x_B^s = x_B^b - \frac{-x_A^b}{RT}\left(\frac{\partial \gamma}{\partial \ln a_B^b}\right)_T \tag{18}$$

or

$$x_B^s = x_B^b - \left(\frac{\bar{\omega}}{RT}\right)\left[\frac{(1-x_B^b)x_B^b}{1+\frac{\partial \ln \phi_B^b}{\partial \ln x_B^b}}\right]\left(\frac{\partial \gamma}{\partial x_B^b}\right)_T \tag{19}$$

where $\bar{\omega}$ denotes A/N_s and ϕ_B^b is the activity coefficient of the B component in the bulk of the liquid alloy.

Laty et al [59] have determined the surface composition of the Al-Cu liquid alloy using Eq. 18, and Saito et al [68] have used Eq. 19 in determining the surface composition of the In-Sn liquid alloy.

Although the equations shown above are modelless, we can utilize adsorption models to derive equations representing the relation between the surface composition and the surface tension. For instance, the most simple ideal monolayer adsorption model [69] gives

$$\frac{x_B^s}{x_A^s} = \frac{x_B^b}{x_A^b} \exp\left[\frac{(\gamma_A^0 - \gamma_B^0)\bar{\omega}}{RT}\right] \tag{20}$$

where γ_j^0 is the pure component surface tension of j and $\bar{\omega}$ is the average of the partial molar surface areas of A and B.

Equation 20 has been applied to study the surface segregation of the Pt-based binary liquid alloy [63] and to study the effects of addition of Cd into Al(liq.) [70]. However, the results are reported to be unsatisfactory.

When the relation between the surface tension and the bulk composition exhibits a complicated curve, as exemplified in

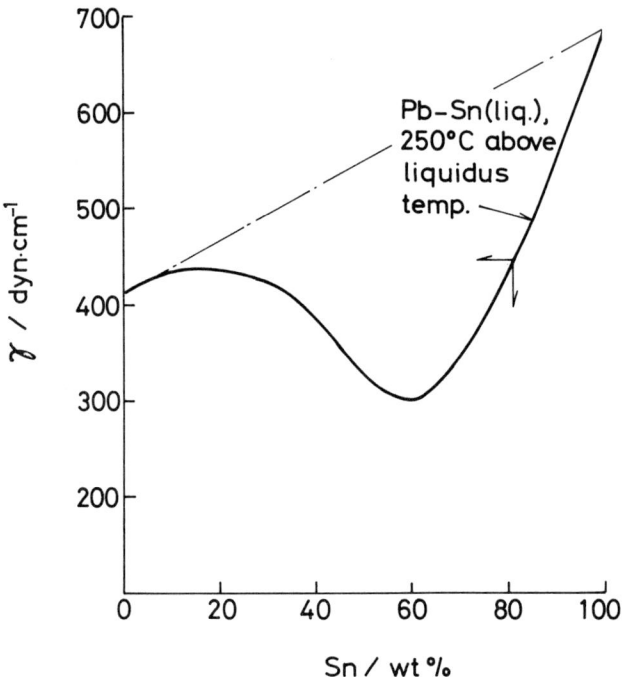

Fig. 1 An anomalous relation between the surface tension and the bulk composition of Pb-Sn liquid alloy [60].

Fig. 1 [60], strong interactions between the atoms constituting the alloy are suggested [71,72]. In such a case, we have to use much more complicated equations based on realistic atomic (molecular) interaction models to calculate the surface composition. Among theoretical equations reported earlier, the regular solution monolayer equation is notable and is frequently reviewed [3,69]. More advanced theoretical studies on this problem have been reported recently [73-76]. These theories involve rather lengthy mathematical treatments, and it is not pertinent to describe them here. The same is the case of the temperature dependence of the surface tension [77].

3. Adsorption

In the preceding discussion on the surface segregation, we have treated the adsorption of atoms that strike the surface from the inside of metallic phase. This section, on the contrary, treats the adsorption in which the adsorbate molecules strike the surface from the outside of the metallic phase.

Now let us consider a gas (or vapor) in contact with a liquid metal. In this case also, we can expect that the adsorption of the gas would bring about the change in the surface tension of the liquid metal because the Gibbs adsorption equation is not restricted by the direction from which the adsorbate molecules approach the surface. Thus we can evaluate the amount of adsorption with Eq. 15: Under ordinary conditions, the activity a_j of the gas is well approximated by the partial pressure p_j, and the surface excess differs little from the absolute amount of adsorption.

Kasama et al [43] have evaluated the amount of oxygen adsorbed by Fe(liq.) using Eq.]5. According to their calculation, the co-area of oxygen is 8.65×10^{-16} cm^2. On the basis of this calculation, they have suggested that the surface is saturated by oxygen and can be regarded as a two-dimensional FeO with a NaCl-type crystal structure. Similarly, amounts of surface impurities such as O, Se, S, and Te on the Fe(liq.) have been estimated using the Gibbs adsorption equation [78]. From the values of co-area, the impurity elements mentioned above are inferred to exist in the ionic states.

When an adsorption takes place onto a liquid alloy, we have to consider the surface segregation together with the adsorption. The basic thermodynamic equation for this system is as follows

$$d\gamma = -\Gamma_A d\mu_A^b - \Gamma_B d\mu_B^b - \Gamma_{gas} d\mu_{gas}^b \qquad (21)$$

If we make the measurement under a constant pressure of the adsorbate gas, it is apparent that $d\mu_{gas}^b = 0$. Thus, the problem formally becomes equivalent to the case of the surface segregation. Utilizing this result, Kishimoto et al [79] have measured the surface segragation of the Ti-Fe liquid alloy under a hydrogen atmosphere. From the theoretical analysis of the experimental data, they have inferred that the arrangement of surface Ti atoms would be 2 × 2 when the concentration of Ti is about 1 atom% while it would be C(2 × 2) when the concentration of Ti is 33.3 atom%. The surface Ti atoms are considered to provide the hydrogen with adsorption sites.

An appropriate assumption for the adsorption isotherm enables us to make further discussion. As an example, let us examine the Langmuir adsorption isotherm

$$\frac{\theta_j}{1 - \theta_j} = K_j a_j \qquad (22)$$

where θ_j is the surface coverage of the adsorbate j, and K_j is the adsorption equilibrium constant for j. By combining Eq. 22 with the Gibbs adsorption isotherm, we obtain

$$\gamma_0 - \gamma = RT\tilde{\Gamma}_j^0 \ln(1 + K_j a_j) \qquad (23)$$

where $\tilde{\Gamma}_j^0$ is the monolayer capacity of adsorption of the surface and γ_0 is the surface tension of a liquid metal or a liquid alloy in an inert atmosphere.

Utilizing the relation mentioned above, Belton [80] has analyzed the surface tension data for many systems: Fe(liq.)-S, Fe(liq.)-S-C, Cu(liq.)-S, Ag(liq.)-O_2, Fe(liq.)-O_2, and Fe(liq.)-Se.

In connection with the applicability of Eq. 23, it must be pointed out that applying this equation to the surface tension data obtained at different temperatures enables us to evaluate the heat of adsorption. According to Belton [80], the heat of adsorption for the Cu(liq.)-S system is ~40 kcal/mol and that for the Ag(liq.)-O_2 system is ~48 kcal/mol.

Problems Related to Surface Tension

An application of an adsorption equation other than the Langmuir equation has been examined by Kemball and Rideal [81,82] in studying the adsorption of various organic molecules onto mercury. According to their report, the Langmuir adsorption equation is not appropriate, but, instead, the following Volmer-type adsorption equation is appropriate for describing the experimental results

$$\Pi(A - b) = kT \tag{24}$$

where Π is the surface pressure defined by $\Pi = \gamma_0 - \gamma$, A is the surface area available for one adsorbate molecule, and b is the co-area of the adsorbate molecule.

It is apparent that $A = b$ when the surface is saturated by the monolayer admolecules. In addition, it must be noted that a good application of Eq. 24 to any adsorption system informs us of an occurrence of a mobile adsorption.

Using an equation obtained by combining Eq. 24 with the Gibbs adsorption isotherm, Kemball has evaluated the amount of adsorption and then determined the enthalpy of adsorption as well as the entropy of adsorption together with the co-area. The thermodynamic data thus obtained have been analyzed using the statistical thermodynamics. Summarized in Table 4 are the results of the analyses mentioned above. In the monolayer region, every adsorbate behaves like a two-dimensional gas. However, in the multilayer region, adsorbed states vary depending on the molecular structures of the adsorbates.

Finally, an adsorption system consisting of a liquid solution and a liquid metal should be discussed. This system is not so familiar to us, but it can nevertheless be treated in the framework of the Gibbs adsorption isotherm.

Ambwami et al [83,84] have measured the surface tension of mercury in contact with a binary solution of an organic reagent and cyclohexane: The organic substance is a solute and cyclohexane is a solvent. It has been found that octadecane as well

TABLE 4 Adsorbed States of Various Organic Molecules on Mercury

	Adsorbed state	
Adsorbate	Monolayer low-coverage region	Multilayer high-coverage region
Benzene	Two-dimensional gas; the plane of benzene ring is parallel to the surface; molecular rotation is allowed only in the plane including benzene ring.	--
Toluene	Two-dimensional gas; the surface mobility is restricted by the methyl group.	--
n-Heptane	Two-dimensional gas; the molecule is adsorbed with 4-5 carbon atoms and part of the molecule is curled up.	--
Methanol	Two-dimensional gas.	
Ethanol	Two-dimensional gas.	
n-Propanol	Two-dimensional gas.	The molecule is too long to form second layer but too short to form condensed phase.
n-Butanol n-Amyl alcohol n-Hexanol	Two-dimensional gas; molecules are lying flat to the surface.	Condensation to liquid film; molecules are erect on the surface.
Acetone	Two-dimensional gas with weak vibration perpendicular to the surface.	Multilayer formation.

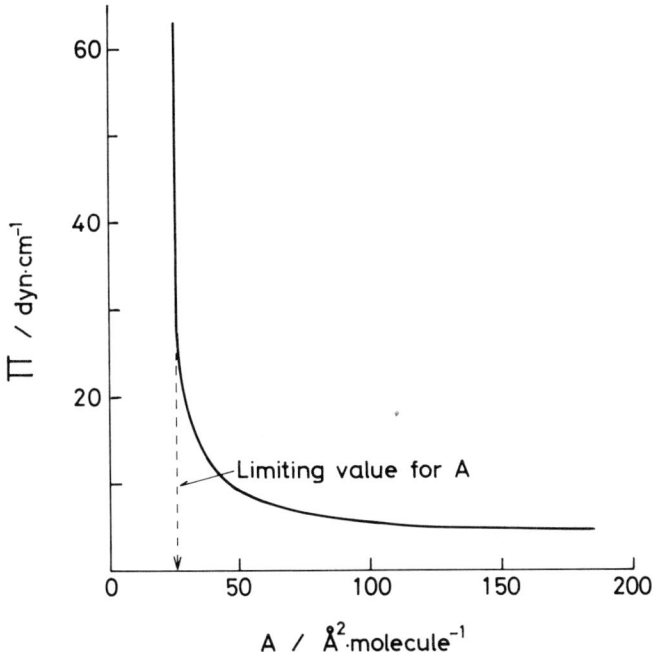

Fig. 2 The relation between the molecular surface area A and the surface pressure Π for the octadecyl alcohol-Hg(liq.) adsorption system at 30°C [84].

as octadecylamine are not adsorbed, while both octadecyl alcohol and stearic acid are adsorbed. According to the result of the thermodynamic analysis of the adsorption data, the entropy of adsorption is positive, suggesting that the adsorption of solute molecules destroys an ordered surface layer of solvent molecules. Further interesting information about the adsorbed phase can be obtained from the adsorption isotherm shown in Fig. 2. As we can see in the figure, the surface pressure Π is small when the surface is sparsely covered by the solute molecules and hence the molecular surface area A is large. On the other hand, the surface pressure steeply increases when the molecular surface area is decreased by the increase in the surface coverage. From this A - Π relation, we can easily estimate the magnitude of the co-area b, i.e., b = ~20-30 $Å^2$. This value of the co-area strongly suggests that the long admolecules are standing vertically to the mercury surface.

E. Correlation of Surface Tension with Other Properties

A correlation of the surface tension (γ) or its temperature derivative ($d\gamma/dT$) with other properties is important from the practical point of view, i.e., an estimation of the surface tension of a given element under any desired conditions. Thus, abundant information about this problem is available in literature.

Usually, the critical temperature T_c is chosen as a quantity to be correlated with the surface tension. White [45] has made a review on this correlation method. Soda et al [34] and Harrison et al [35] have published results of correlation between T_c and $d\gamma/dT$ for Cu(liq.). It is apparent that exact values of T_c are indispensable to making the correlation reliable. However, the experimental determination of the exact T_c value is considered to be more difficult than the surface tension measurement. For this reason, Lang [55] and Blairs [85] have found equations capable of evaluating the critical temperature of any desired element. After knowing the value of T_c, we can easily make an estimation of γ or of $d\gamma/dT$: The equations necessary for this estimation are seen in the review by White [45]. The values of $d\gamma/dT$ estimated by Lang [55] are listed in Table 2 in order to facilitate the comparison with the values reported by other workers.

Correlations of the surface tension with bulk properties other than the critical temperature have also been examined. Miedema and Boom [86] have found that the surface tension (γ_m) at the melting point can be correlated with $\Delta H_{0,v}/V_m^{2/3}$, where $\Delta H_{0,v}$ is the heat of vaporization at zero temperature, and V_m is the atomic volume at the melting point. Values of the heat of vaporization necessary for the correlation are tabulated by Miedema. Papazian [87] has proposed another method of correlation for γ_m. Namely, he has shown that γ_m can be correlated with the room temperature bulk modulus (β) of a metal by $\gamma_m = C\beta^{2/3}$, where C is a

Problems Related to Surface Tension

proportionality constant. Another method reported by Lang [88] is necessary for the correlation are tabulated by Miedema. Papazian [87] has proposed another method of correlation for γ_m. Namely, he has shown that γ_m can be correlated with the room temperature bulk modulus (β) of a metal by $\gamma_m = \tilde{C}\beta^{2/3}$, where $\tilde{C}$ is a proportionality constant. Another method reported by Lang [88] is capable of estimating the values of γ_m for all elements up to an atomic number of 95.

The temperature dependence of the surface tension has also been correlated with the bulk properties other than the critical temperature. Namely, Papazian [56] has shown that the thermal expansion can be correlated with $d\gamma/dT$. The estimated values of $d\gamma/dT$ are shown in Table 2 [56] in order to compare with other reported values.

Generally speaking, the correlation methods mentioned above are mostly empirical and the underlying models are vague. In contrast to this Iida et al [89] have proposed a model as shown in Fig. 3. They have postulated that the work W necessary for separating the liquid metal into two parts A and B by a dividing plane XX' would be related to the surface tension. According to this view, the surface tension is given by the following equation

$$\gamma = \frac{1}{2} W = \left(\frac{1}{4}\right) \rho \bar{r} w z \qquad (25)$$

where ρ is the density of the liquid metal, $\bar{r}$ is the mean atomic distance, z is the coordination number, and w is the energy necessary for separating the atom 1 from the atom 2 along the direction YY' perpendicular to the dividing plane XX'. By combining Eq. 25 with the Lindemann's theory of melting [90], Iida has derived the following relation:

$$\gamma_m = \frac{4T_m}{V_m^{2/3}} \qquad (26)$$

Finally we have to pay attention to the correlation of the surface tension with electronic properties of the metal as well

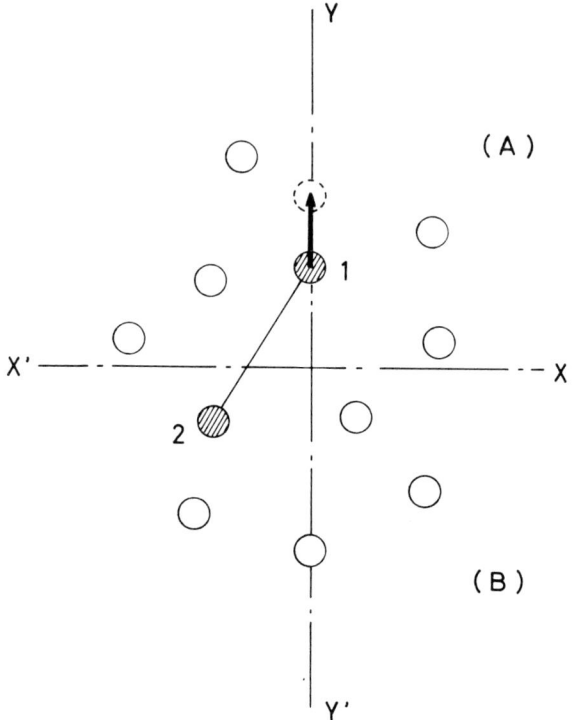

Fig. 3 An atomic model for the origin of the surface tension of the liquid metal [89].

[86]. According to Papazian [91], it is possible to correlate γ_m with the plasmon frequency. This kind of approach is not theoretically improbable because the larger part of the surface tension of the liquid metal is attributed to electrons.

REFERENCES

1. T. E. Faber, *Theory of Liquid Metals*, Cambridge University Press, 1972.
2. S. Z. Beer, ed., *Liquid Metals Chemistry and Physics*, Marcel Dekker, Inc., New York, 1972.
3. M. Shimoji, *Liquid Metals*, Academic Press, London, New York, San Francisco, 1977.

4. B. C. Allen, in Ref. 2, p. 161 (1972).
5. Ref. 6, p. 277 (1960).
6. S. Ono and S. Kondo, in *Handbuch der Physik,* vol. X, sect. 1, Springer-Verlag, Berlin, Gottingen, Heidelberg, 1960, p. 134.
7. C. A. Croxton and R. P. Ferrier, *J. Phys. C. Solid St. Phys., 4,* 1909 (1971).
8. C. A. Croxton, *Adv. Phys., 22,* 385 (1973).
9. F. F. Abraham, *Phys. Rep., 53,* 93 (1979).
10. R. Evans, *Adv. Phys., 28,* 143 (1979).
11. V. K. Semenchenko, *Surface Tension in Metals and Alloys,* Pergamon Press, Oxford, 1961.
12. C. A. Croxton, *Statistical Mechanics of the Liquid Surface,* John Wiley & Sons, Chichester, New York, Brisbane, Toronto, 1980.
13. R. H. Fowler, *Proc. Roy. Soc. London, A159,* 229 (1937).
14. Y. Waseda and M. Ohtani, *Z. Naturforsch., 30a,* 485 (1975).
15. L. V. Belan-Gayko, V. I. Bogdanov, and D. L. Fuks, *Phys. Met. Metall., 47,* 195 (1980).
16. B. M. Lepinskikh and N. V. Stepanova, *Russ. J. Phys. Chem., 54,* 452 (1980).
17. R. Kumaravadivel and R. Evans, *J. Phys. C: Solid St. Phys., 8,* 793 (1975).
18. R. Evans and R. Kumaravadivel, *J. Phys. C: Solid St. Phys., 9,* 1891 (1976).
19. K. K. Mon and D. Storch, *Phys. Rev. Lett., 45,* 817 (1980).
20. M. Hasegawa, M. Watabe, and W. H. Young, *J. Phys. F: Metal Phys., 11,* L173 (1981).
21. R. Evans and M. Hasegawa, *J. Phys. C: Solid St. Phys., 14,* 5225 (1981).
22. M. Hasegawa and M. Watabe, *J. Phys. C: Solid St. Phys., 15,* 353 (1982).
23. E. Chacon, F. Flores, and G. Navascues, *J. Phys. C: Solid St. Phys., 16,* L187 (1983).
24. M. Hasegawa and M. Watabe, *J. Phys. C: Solid St. Phys., 16,* L669 (1983).
25. E. Chacon, F. Flores, and G. Navascues, *J. Phys. C: Solid St. Phys., 16,* L701 (1983).
26. J. Goodisman and M-L. Rosinberg, *J. Phys. C: Solid St. Phys., 16,* 1143 (1983).

27. D. M. Wood and D. Stroud, *Phys. Rev. B*, *28*, 4374 (1983).
28. S. A. Trigger, *Solid State Comm.*, *52*, 391 (1984).
29. D. Stroud and M. J. Grimson, *J. Non-Crystall. Sol.*, *68*, 231 (1984).
30. S. M. Foiles and N. W. Ashcroft, *Phys. Rev. A*, *30*, 3136 (1984).
31. R. Evans, *J. Phys. C: Solid St. Phys.*, *7*, 2808 (1974).
32. E. Chacon, F. Flores, and G. Navascues, *J. Phys. F: Met. Phys.*, *14*, 1587 (1984).
33. A. Kasama, T. Iida, and Z. Morita, *J. Jpn. Inst. Metals*, *40*, 1030 (1976).
34. H. Soda, A. McLean, and W. A. Miller, *Trans. Jpn. Inst. Metals*, *18*, 445 (1977).
35. D. A. Harrison, D. Yan, and S. Blairs, *J. Chem. Thermodynamics*, *9*, 1111 (1977).
36. G. Lang, P. Laty, J. C. Joud, and P. Desre, *Z. Metallkde*, *68*, 113 (1977).
37. B. B. Alchagirov, S. N. Zadumkin, M. B. Kokov, B. K. Unezhev, and K. B. Khokonov, *Russ. Met.*, 73 (1979).
38. L. Goumiri, J. C. Joud, P. Desre, and J. M. Hicter, *Surf. Sci.*, *83*, 471 (1979).
39. K. B. Khokonov, S. N. Zadumkin, and B. B. Alchagirov, *Elektrokhimiya*, *10*, 911 (1974).
40. G. J. Abbachian, *J. Less-Common Met.*, *40*, 329 (1975).
41. G. Lang, *J. Inst. Metals*, *101*, 300 (1972).
42. A. Passerone, R. Sangiorgi, and G. Caracciolo, *J. Chem. Thermodynamics*, *15*, 971 (1983).
43. A. Kasama, A. McLean, W. A. Miller, Z. Morita, and M. J.
44. U. Mittag and K. W. Lange, *Arch. Eisenhuttenwes.*, *46*, 249 (1975).
45. D. W. G. White, *Metall. Rev.*, *13*, 73 (1968).
46. I. S. Kisil, A. G. Malko, and M. M. Dranchuk, *Russ. J. Phys. Chem.*, *55*, 177 (1981).
47. B. T. Belov, *Russ. J. Phys. Chem.*, *55*, 302 (1981).
48. A. Pamies, C. G. Cordovilla, and E. Louis, *Scr. Metall.*, *18*, 869 (1984).
49. Z. Morita and A. Kasama, *Trans. Jpn. Inst. Metals*, *21*, 522 (1980).

References

50. R. C. Brown and N. H. March, *J. Phys. C: Solid St. Phys.*, *6*, L363 (1973).
51. Y. Waseda and K. T. Jacob, *Phys. Stat. Sol.*, *(a)68*, K117 (1981).
52. T. Itami and M. Shimoji, *J. Phys. F: Metal Phys.*, *9*, L15 (1979).
53. W. F. Carnahan and K. E. Starling, *J. Chem. Phys.*, *51*, 635 (1969).
54. B. N. Tewari, B. S. Bhargava, and K. N. Khanna, *Phys. Stat. Sol.*, *(b)114*, K31 (1982).
55. G. Lang, *Z. Metallkde*, *68*, 213 (1977).
56. H. A. Papazian, *Scr. Metall.*, *18*, 1401 (1984).
57. A. Adachi, Z. Morita, Y. Kita, A. kasama, and S. Hamamatsu, *J. Jpn. Inst. Metals*, *35*, 1188 (1971).
58. A. E. Schwaneke, W. L. Falke, and V. R. Miller, *J. Chem. Eng. Data*, *23*, 298 (1978).
59. P. Laty, J. C. Joud, P. Desre, and G. Lang, *Surf. Sci.*, *69*, 508 (1977).
60. M. Demeri, M. Farag, and J. Heasley, *J. Mater. Sci.*, *9*, 683 (1974).
61. K. Okajima and H. Sakao, *Trans. Jpn. Inst. Metals*, *23*, 111 (1982).
62. V. M. Chumarev, V. M. Sholokhov, A. I. Okunev, and V. P. Chentsov, *Russ. Metall.*, 27 (1980).
63. A. I. Chegodayev, E. L. Dubinin, M. M. Mitko, and A. I. Timofeyer, *Russ. Metall.*, 64 (1979).
64. V. S. Savvin and K. I. Ibragimov, *Russ. Metall.*, 59 (1977).
65. B. Galloisand and C. H. P. Lupis, *Metall. Trans. B*, *12B*, 679 (1981).
66. L. L. Migai, N. Y. Mikhailov, N. L. Perlova, M. A. Pokrasin, V. V. Roshchupkin, and A. I. Chernov, *Russ. J. Phys. Chem.*, *55*, 1539 (1981).
67. H. H. Liebermann, *J. Mater. Sci.*, *19*, 1391 (1984).
68. Y. Saito, H. Yoshida, T. Yokoyama, and Y. Ogino, *J. Colloid. Interf. Sci.*, *66*, 440 (1978).
69. S. H. Oberburg, P. A. Bertrand, and G. A. Somorjai, *Chem. Rev.*, *75*, 547 (1975).
70. P. S. Popel, B. P. Domashnikov, V. M. Zamyatin, Y. A. Bazin, B. A. Baum, V. A. Konovalov, and E. E. Baryshev, *Russ. Metall.*, 49 (1983).

71. P. R. Couchman and C. L. Reynolds, Jr., *J. Mater. Sci.*, *10*, 1089 (1975).
72. P. S. Popel, V. M. Zamyatin, B. P. Domashnikov, Y. A. Bazin, V. A. Konovalov, V. A. Pavlov, and Y. A. Nasyyrov, *Russ. Metall.*, 38 (1983).
73. F. L. Williams and D. Nason, *Surf. Sci.*, *45*, 377 (1974).
74. L. Goumiri, J. C. Joud, and P. Desre, *Surf. Sci.*, *88*, 461 (1979).
75. A. L. Bolshov, *Russ. J. Phys. Chem.*, *55*, 438 (1981).
76. L. Goumiri and J. C. Joud, *Surf. Sci.*, *138*, 524 (1984).
77. J. C. Joud, N. Eustathopoulos, and P. Desre, *J. Chim. Physiq.*, *71*, 559 (1974).
78. K. Ogino, K. Nogi, and O. Yamase, *Trans. ISIJ.*, *23*, 234 (1983).
79. M. Kishimoto, K. Mori, and Y. Kawai, *J. Jpn. Inst. Metals*, *48*, 413 (1984).
80. G. R. Belton, *Metall. Trans. B*, *7B*, 35 (1976).
81. C. Kemball and E. K. Rideal, *Proc. Roy. Soc.*, *A187*, 53 (1946).
82. C. Kemball, *Proc. Roy. Soc.*, *A187*, 73 (1946); *A190*, 117 (1947).
83. D. S. Ambwami and T. Fort, Jr., *J. Colloid Interf. Sci.*, *42*, 1 (1973).
84. D. S. Ambwami, R. A. Jao, and T. Fort, Jr., *J. Colloid Interf. Sci.*, *42*, 8 (1973).
85. S. Blairs, *J. Colloid Interf. Sci.*, *67*, 548 (1978).
86. A. R. Miedema and R. Boom, *Z. Metallkde*, *69*, 183 (1978).
87. H. A. Papazian, *High Temp. Sci.*, *18*, 53 (1984).
88. G. Lang, *Z. Metallkde*, *67*, 549 (1976).
89. T. Iida, A. Kasama, M. Misawa, and Z. Morita, *J. Jpn. Inst. Metals*, *38*, 177 (1974).
90. F. A. Lindemann, *Phys. Z.*, *11*, 609 (1910).
91. H. A. Papazian, *High Temp. Sci.*, *18*, 19 (1984).

6
Optical Properties and Surface Transition Zone

I.	Introduction	155
II.	Methods of Studying Optical Properties of Liquid Metals	156
	A. Relations Among Optical Constants in Drude Theory	156
	B. Ellipsometry and Reflectometry	157
III.	Overview of Experimental Results	163
IV.	Surface Transition Zone	168
	A. Anomalous Properties of Hg(liq.) Surface	168
	B. Advanced Studies on Transition Zone	171
	C. Comments on the Transition Zone with Special Structure	176
	References	177

I. INTRODUCTION

A number of papers reporting optical properties of liquid metals have been published. Since ordinary electromagnetic waves penetrate to the depth of 130-500 Å from the surface of a liquid metal [1], the optical properties measured by usual methods provide us with information about the electronic states in the bulk phase. Thus, most existing reviews [2,3] are concerned mainly with the relation between optical properties and bulk electronic properties of liquid metals.

The purpose of this chapter differs from those of other reviews—it presents information about the relation between the optical properties and the surface properties of liquid metals. However, as we can see in the text, traditional optical studies of liquid metals are the sources of the optical studies of the liquid metal surfaces. Therefore we have to begin with the understanding of a few theoretical and experimental bases of traditional optical studies.

II. METHODS OF STUDYING OPTICAL PROPERTIES OF LIQUID METALS

A. Relations Among Optical Constants in Drude Theory

Optical properties of a substance are usually expressed by its dielectric constant ε or its refractive index n. These are both complex quantities which are usually expressed by the following relations

$$\begin{aligned} \varepsilon(\omega) &= \varepsilon_1(\omega) + i\varepsilon_2(\omega) \\ n(\omega) &= n_1(\omega) + in_2(\omega) \end{aligned} \tag{1}$$

where ω is the angular frequency of the electromagnetic wave and the subscripts 1 and 2 denote the real part and the imaginary part of the quantities subscripted, respectively.

According to the theory of Drude, the optical properties of a simple metal are determined by an interaction between the electromagnetic wave and electrons in the metal and the following relations between ε and n are obtainable

$$\varepsilon_1 = n_1^2(\omega) - n_2^2(\omega) = 1 - \frac{4\pi n^* e^2 \tau^2}{m(1 + \omega^2\tau^2)} = 1 - \frac{\omega_p^2 \tau^2}{1 + \omega^2\tau^2} \tag{2}$$

$$\varepsilon_2 = 2n_1(\omega)n_2(\omega) = \frac{4\pi n^* e^2 \tau}{n\omega(1 + \omega^2\tau^2)} = \frac{\omega_p^2 \tau}{\omega(1 + \omega^2\tau^2)} \tag{3}$$

where $-e$ is the electronic charge, m is the mass of electron, n* is the electron density, τ is the relaxation time of conduction, and ω_p, which is defined by $\omega_p = (4\pi n^* e^2/m)^{1/2}$, is the plasma frequency of free electrons.

In addition, the real part σ_1 and the imaginary part σ_2 of the optical conductivity $\sigma(\omega)$ can be related to the direct current conductivity (σ_0) by the following relations

$$\sigma_1(\omega) = \frac{\sigma_0}{1 + \omega^2 \tau^2} \qquad (4)$$

$$\sigma_2(\omega) = \frac{\sigma_0 \omega \tau}{1 + \omega^2 \tau^2} \qquad (5)$$

The relations mentioned above provide us with theoretical bases in studying electronic structures of liquid metals by means of optical measurements. If optical properties of a liquid metal obey the Drude theory, we can infer that the electronic properties of this metal would be described by the free electron theory of liquid metals. On the other hand, if this is not the case, we have to find any appropriate reasons. It must be pointed out that an anomalous surface structure can be the cause of the deviation of the optical properties from simple free electron theory.

B. Ellipsometry and Reflectometry

1. *Ellipsometry*

The use of the ellipsometer enables us to measure two quantities Ψ and Δ, which are known as the restored azimuth and as the differential phase change on reflection, respectively. These two quantities are related to the complex reflection amplitude (r_p) for the p-polarized light and that (r_s) for the s-polarized light (ref. Fig. 1) through the following relation

$$\frac{r_p}{r_s} = \frac{|r_p| \exp(i\delta_p)}{|r_s| \exp(i\delta_s)} = \left(\frac{|r_p|}{|r_s|}\right) \exp[i(\delta_p - \delta_s)] = \tan \Psi \exp(i\Delta) \qquad (6)$$

where $|r_p|/|r_s| \equiv \tan \Psi$, and $\delta_p - \delta_s \equiv \Delta$.

It is known that the restored azimuth Ψ and the phase change Δ can be related to the dielectric constant $\varepsilon_1(\omega)$ and $\varepsilon_2(\omega)$ by the following relations [4]

$$\varepsilon_1 = \sin^2\varphi \, \tan^2\varphi \, \frac{\cos^2 2\Psi - \sin^2 2\Psi \sin^2\Delta}{(1 + \cos \Delta \sin 2\Psi)^2} + \sin^2\varphi \qquad (7)$$

$$\varepsilon_2 = 2 \sin^2\varphi \, \tan^2\varphi \, \frac{\sin 2\Psi \cos 2\Psi \sin \Delta}{(1 + \cos \Delta \sin 2\Psi)^2} \qquad (8)$$

where φ is the angle of incidence.

The optical cell which serves for the ellipsometric measurement has to be specially designed to protect the liquid metal from contaminations. In ordinary measurements, therefore, the optical cell is either evacuated or filled with purified hydrogen [4-6]. A glass-made optical cell designed by Faber and Smith is shown in Fig. 2: The electrode serves for eliminating contaminations from the sample surface by means of an electric discharge. In special cases, e.g., measurements in the vacuum ultra-violet

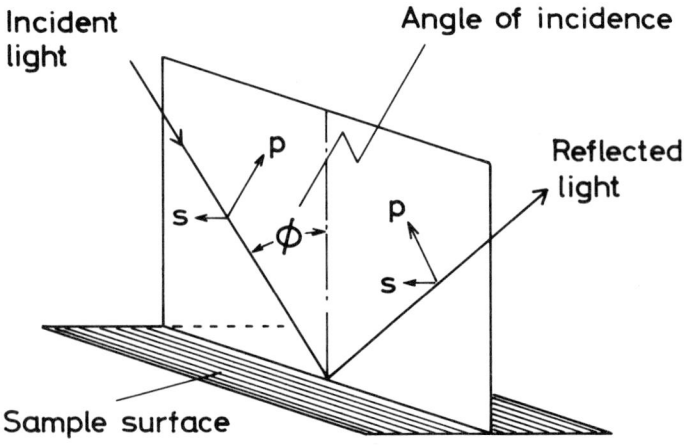

Fig. 1 An incidence and a reflection of a light beam on and from the liquid metal surface.

Optical Properties of Liquid Metals 159

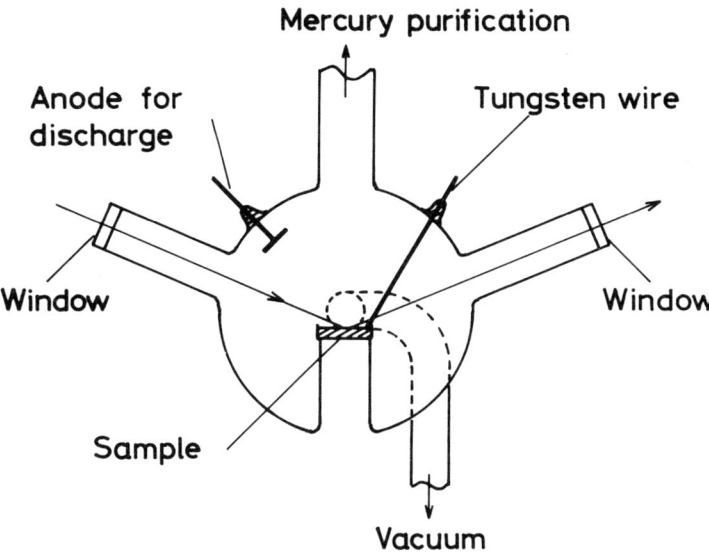

Fig. 2 An ellipsometric optical cell [4].

region, almost whole optical assemblies are enclosed in a vacuum chamber [7]. For most liquid metals, sample holders are made from alumina [5,6] whereas the following metallic sample holders are recommended by some authors: molybdenum for Cu(liq.) [5,6], steel for Ag(liq.) [5], and tungsten for Ga(liq.) and Sn(liq.) [6]. It is reported that the use of the above-mentioned metallic sample holders enables us to keep the meniscus in a flat state.

2. Reflectometry

There are two basic equations that relate the complex reflection amplitudes (r_p and r_s) to the angle of incidence φ

$$r_p = \frac{n_0^2(n^2 - n_0^2 \sin^2\varphi)^{1/2} - n^2(n_0^2 - n_0^2 \sin^2\varphi)^{1/2}}{n_0^2(n^2 - n_0^2 \sin^2\varphi)^{1/2} + n^2(n_0^2 - n_0^2 \sin^2\varphi)^{1/2}} \quad (9)$$

$$r_s = \frac{(n_0^2 - n_0^2 \sin^2\varphi)^{1/2} - (n^2 - n_0^2 \sin^2\varphi)^{1/2}}{(n_0^2 - n_0^2 \sin^2\varphi)^{1/2} + (n^2 - n_0^2 \sin^2\varphi)^{1/2}} \quad (10)$$

where n_0 is the complex refractive index of the dielectrics in contact with the sample surface: $n_0 = n_{0,1} + in_{0,2}$, where $n_{0,1}$ and $n_{0,2}$ are the real part and the imaginary part, respectively.

Unfortunately, however, we can not determine the magnitude of n directly from Eqs. 9 and 10 and experimental data because the quantities that can be measured by the reflectometry are $R_p = r_p r_p^*$ and $R_s = r_s r_s^*$ (the superscript * denotes a complex conjugate) [8]. Thus, we have to evaluate n_1 and n_2 through various combinations of R_p and R_s [7-9].

There is another way of using Eqs. 9 and 10 in determining the magnitudes of optical constants. The method is called the normal incidence method and employs a special relation derived from Eqs. 9 and 10 at $\varphi = 0$, i.e.,

$$R_p = R_s \equiv R_N = \frac{(n_1 - n_{0,1})^2 + (n_2 - n_{0,2})^2}{(n_1 + n_{0,1})^2 + (n_2 + n_{0,2})^2} \tag{11}$$

This equation is much more simple compared with Eqs. 9 and 10 and appears convenient to use. Indeed, as we can see later, a number of optical studies of liquid metals have been carried out by the normal reflection method. However, it is necessary to recognize that the normal reflection method involves a few difficulties.

The first difficulty is that we cannot rigorously realize the conditions of the normal reflection because it is impossible to put the light source, the sample, and the detector on one optical axis. Thus, the angle of incidence of ~10° is adopted in usual optical studies. Fortunately, in the visible and infrared regions, the error introduced by this approximation is reported to be insignificant [8,10].

The second difficulty is that any single measurement of R_N is insufficient to determine the optical constants of the sample because Eq. 11 contains two unknown quantities n_1 and n_2. We have to make at least two independent measurements by changing the dielectrics, i.e., by changing the values of $n_{0,1}$ and $n_{0,2}$. Bloch and Rice [8] have measured the optical constants of Hg-(liq.) by means of the normal reflection method using four different dielectricses as windows in contact with the sample surface: lithium fluoride, sodium chloride, cesium bromide, and KRS-5 (thallium bromide-iodide).

An alternative way to supplement the insufficient information obtained by a single measurement of R_N is to use the following relations, which are called the Kramers-Kronig relations [8,10,11]

$$\ln r(\omega) = \frac{2}{\pi} \int_0^\infty \frac{\omega'\delta(\omega') - \omega\delta(\omega)}{\omega'^2 - \omega^2} d\omega' \tag{12}$$

$$\delta(\omega) = \frac{2\omega}{\pi} \int_0^\infty \frac{\ln r(\omega') - \ln r(\omega)}{\omega'^2 - \omega^2} d\omega' \tag{13}$$

where r and δ are the amplitude and the phase, which are defined by the complex amplitude reflection coefficient at normal incidence, i.e., $r_N = r \exp(i\delta)$.

The normal reflection measurement gives $r(\omega)$ because $r = R_N^{1/2}$. Thus, if we have enough data for $r(\omega)$ over a wide frequency range to make accurate extrapolation to $\omega' \to 0$ and to $\omega' \to \infty$, we can evaluate $\delta(\omega)$ from Eq. 13, and then we can obtain n_1 and n_2 from

$$r \exp(i\delta) = \frac{n_1 - in_2 - n_0}{n_1 - in_2 + n_0} \tag{14}$$

In practice, it is not always possible to obtain $r(\omega)$ over a wide frequency range. Thus this method is available only in limited cases [8,10,11].

Illustrated in Fig. 3 are several designs of optical cells used in the reflectometry. Among these, Fig. 3a [12], Fig. $3b_1$ [7], and Fig. $3b_2$ [13] are available for the reflectivity measurements at oblique incidences. Hodgson has reported that the

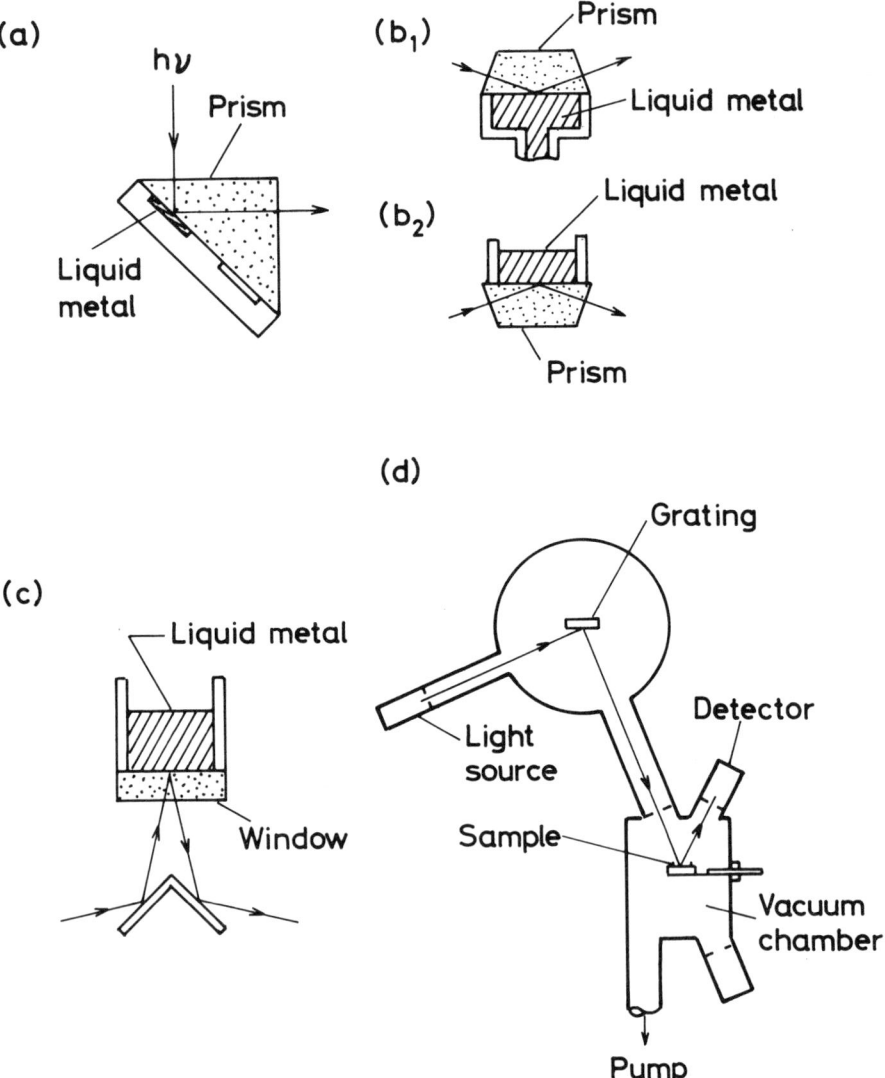

Fig. 3 Several reflectometric optical cells of different designs [7,8,11-13].

design in Fig. $3b_2$ is not recommended because air film adhering on the prism surface as well as a strain double refraction bring about significant errors. From this reason he has used the free surface for the reflection measurement [13]. The design in Fig. 3c is used for normal reflection measurements [8,11]. The bottom window is exchangeable, and this enables us to take data necessary for evaluating the optical constants using Eq. 11. The design in Fig. 3d is applicable for the normal reflection measurements in the vacuum ultraviolet region [10,11,14]. It is reported that a MgF_2 window floated on the sample surface prevents the surface from oxidation and vibration [11]. In the designs from Fig. 3a-d, cases where a dielectrics (prism or window material) is in contact with the sample surface are not scarce. In such cases we have to correct the experimental results for the effects of the dielectrics [7,8].

Irrespective of the cell designs mentioned above, we have to pay great attention to avoid surface contaminations. For this purpose, an enclosing of main optical units in a vacuum chamber as adopted by Crozier and Murphy [7] appears desirable, although the use of a gas-tight chamber, which permits a continuous flushing of an inert gas, is also effective to a certain extent [8].

III. OVERVIEW OF EXPERIMENTAL RESULTS

Many papers describing the optical properties of liquid metals and liquid alloys have been published since 1970, as we can see in Tables 1 and 2. From a survey of these tables, we can draw the following conclusions:

1. The Drude theory cannot be applied to describe the optical properties of Fe(liq.), Co(liq.), and Ni(liq.).
2. Most of the liquid metals, including noble metals such as Ag(liq.) and Au(liq.), obey the Drude theory in the low-energy region of the incident light: Excitations of core electrons take place at high-energy regions.

TABLE 1 Ellipsometric Studies

Liquid metals and liquid alloys[a]	Wave number (cm^{-1})	Temperature (°C)	Pressure (Torr)	Results[b]	Remarks	Ref.
Hg					Free surface; purification by discharge in Ar (0.1 Torr)	4
Cd	14,000–25,000	--	10^{-5}	ND	Glass-metal; quartz-metal contacts	
Pb						
Bi						
--						
Al	6,200–40,000	900		D^c	Surface contamination; free surface	5
--						
Cu				D^c	Free surface; excitation of electrons from the d-band at a high-energy region	5
Ag	6,200–40,000	Melting point	10^{-1} (H_2)	D^c		
Au				ND		
Fe				ND		
Co				ND	Free surface	5
Ni				ND		
--						
Hg	16,000–24,000	27	--	--	Larger absolute values of ε_1 and ε_2 than those obtained by the reflectometry	15
--						
Pb	1,250–11,000	600	$(10^{-5})^d$	D	--	6

Overview of Experimental Results

Metal	Wavelength range	Temperature	Vacuum	Drude	Comments	Ref.
Bi	1,250–11,000	800	$(10^{-5})^d$	D	--	6
Cu		1100			Free surface	ε
Au		1100				6
Al	1,250–11,000	900, 1170	$(10^{-5})^d$	D	Free surface; surface contaminations are severe for Al(liq.)	6
Ga		600, 860				
Sn		600, 800, 1170			Free surface	6
Hg	1,250–11,000	Room temp.	$(10^{-5})^d$	ND		
Hg	13,000–40,000	22	10^{-6}	ND	Free surface; optical systems are placed in a vacuum chamber	7
Hg	13,300–40,000	22	10^{-6}	ND	The liquid metal is contacted with dielectrics	7
Cu(30)-Sn(70)						
Cu(70)-Sn(30)	1,250–11,000	1150	$(10^{-5})^d$	D		6
Cu(90)-Sn(10)						

[a] Values in parentheses are compositions in atom%.
[b] D, Drude-like; ND, nonDrude-like.
[c] Only crudely Drude-like at a low-energy region.
[d] Best vacuum attainable in the apparatus.

TABLE 2 Reflectometric Studies

Liquid metals and liquid alloys	Wave number (cm^{-1})	Temperature (°C)	Pressure (Torr)	Results[a]	Remarks[b]	Ref.
Ga	770–44,000	30	--	D	Window: glass, quartz, or NaCl	12
Hg	770–44,000	20	--	D		12
Hg	4,000–17,000	20, 230	--	ND	Free surface; in H_2, in air, or in vacuum	13
In	4,000–27,000	450, 813	--	D	Heating in H_2 at 850°C for cleaning	16
Cd	4,000–27,000	426, 450	--	D	Surface contamination	16
Bi	4,000–27,000	505, 695	--	D	Heating in H_2	16
Sb	4,000–27,000	654, 824	--	D	Heating in H_2	16
Hg	16,000–160,000	25	Max 10^{-6}	D	RN(10°)	10
In	16,000–160,000	170	Max 10^{-6}	ND	RN(10°) ⎫ Excitation of core electrons at a high energy region	10
Bi	16,000–160,000	350	Max 10^{-6}	ND	RN(10°) ⎭	10
Hg	333–20,000	--	--	D	RN(<10°); optical assemblies are placed in a gas-tight box and flushed with Ar or N_2	8

Overview of Experimental Results

Material	Range			D/ND	Notes	Ref
Hg	4,000–40,000	Room temp.	--	D	RN or oblique incidence, sapphire (window, prism)	17
Hg	10,000–22,000	25	10^{-5}	D	RN(10°); free surface in vacuum in He	14
Hg	4,000–66,000	27	10^{-6}	--	RN(7°); MgF_2 window	15
Hg	13,300–40,000	22	10^{-6}	D for R_s ND for R_p	Silica–Hg(liq.) system	7
Hg	13,300–40,000	22	10^{-6}	ND for R_s ND for R_p	Borosilicate–Hg(liq.) system	7
Hg-In	2,500–71,400	--	10^{-5}	ND	RN; electronic spectra in 2–9 eV region; sensitive to surface oxidation; MgF floated over Hg(liq.) protects the surface from oxidation and vibration	11
Hg-In	4,000–77,000	20	--	--	--	13

[a] D, Drude-like; ND, nonDrude-like; R_s, reflection for s-polarized light; R_p, reflection for p-polarized light.

[b] RN, Normal incidence (the angle of incidence is shown in parentheses).

3. Mercury, Hg(liq.), is exceptional among simple liquid metals. Results of ellipsometric measurements are mostly nonDrude-like, while those of reflectometric measurements are mostly Drude-like.

IV. SURFACE TRANSITION ZONE

A. Anomalous Properties of Hg(liq.) Surface

The facts that most of the simple liquid metals and simple liquid alloys are Drude-like in their optical behaviors mean that we can approximately regard the electronic structures of these liquids as homogeneous from the surface to the interior of the bulk phase. The sharp surface model (ss-model) illustrated in Fig. 4a represent such a case, as mentioned above [18]. The z-coordinate is defined perpendicular to the surface and the position of the Gibbs dividing plane is defined as the origin: z > 0 denotes the interior of the liquid metal and z < 0 denotes the outside of the liquid metal (vapor phase).

So long as we accept item 3 in the list above, the ss-model cannot be applied to explain the optical properties of Hg(liq.). Bloch and Rice [8] have postulated that the anomalous optical properties of Hg(liq.) would be ascribed to the transition zone existing between the bulk phase and the vapor phase. Thus they have analyzed the experimental data and have shown that both ellipsometric results and reflectometric results can be explained if we assume a surface transition zone. It is assumed that the surface transition zone has an optical conductivity, which is given by

$$\sigma(w) = \left[\frac{w}{(1+w)}\right]\left[\sigma_b + \frac{\sigma_s}{1+w}\right] \qquad (15)$$

where $w = \exp(z/\Delta)$, σ denotes the optical conductivity, Δ is the parameter related to the width of the transition zone (1-5 Å), and the subscripts b and s denote the bulk and surface, respectively.

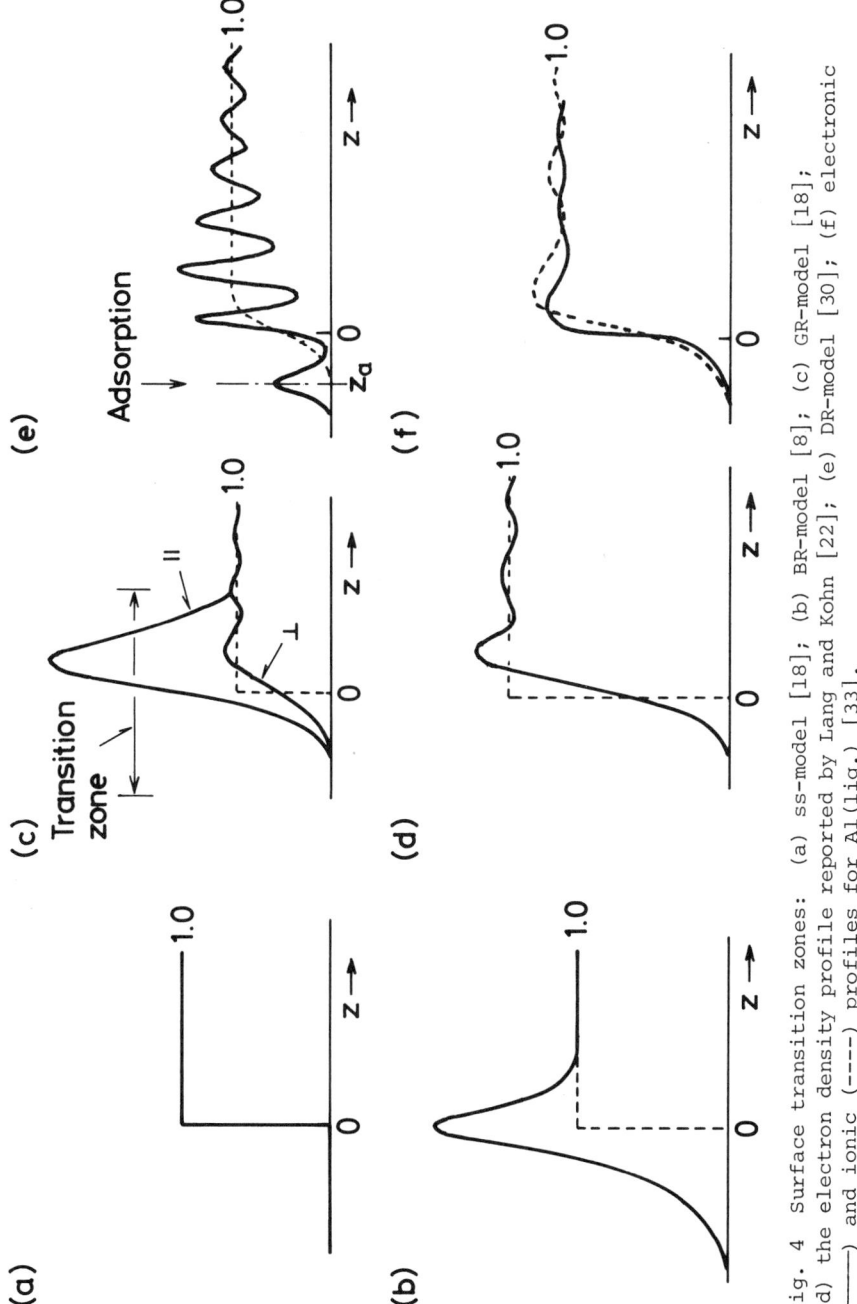

Fig. 4 Surface transition zones: (a) ss-model [18]; (b) BR-model [8]; (c) GR-model [18]; (d) the electron density profile reported by Lang and Kohn [22]; (e) DR-model [30]; (f) electronic (———) and ionic (----) profiles for Al(liq.) [33].

The transition zone which is characterized by Eq. 15 is called the Bloch-Rice model (BR-model) and is shown schematically in Fig. 4b.

In addition to the above-mentioned transition zone model, Bloch and Rice have proposed another model which is called the skewed BR-model. In this model, the transition zone is divided into two parts at the position of the maximum conductivity. It has been shown that the experimental optical properties of Hg-(liq.) can be explained if we assume that the outer width (Δ_1) and the inner width (Δ_2) of the transition zone are 1 Å and 5 Å, respectively.

The BR-model as well as the skewed BR-model have been utilized by other workers and provided them bases of discussion. For instance, Crozier and Murphy [7] have analyzed the optical data of Hg(liq.) using these models and reached the following conclusion. Namely, the surface optical conductivity of Hg(liq.) is vectorial and the electric field applied perpendicular to the surface is sensitive while the field applied parallel to the surface is insensitive to the transition zone. Siskind et al [11] have applied the skewed BR-model to the analysis of the optical data for the In-Hg liquid alloy. In this case, the analysis has been made by taking the surface segregation into account. It has been recognized that the surface of the liquid alloy is enriched by mercury and the concentration gradient is extended to the depth of about 30 Å from the surface. Therefore, the bulk conductivity σ_b is no longer a constant but instead is a function of z. Thus, Siskind has revised the expression for the bulk conductivity in the following manner

$$\sigma_b(z) = \sigma_b(0) \quad z < 0$$
$$\sigma_b(z) = \sigma_b^* + [\sigma_b(0) - \sigma_b^*] \exp(-\frac{z}{\Lambda}) \quad (16)$$

where σ_b^* denotes the optical conductivity of the bulk where the concentration gradient has disappeared, and Λ is a parameter representing the depth of the surface segregation zone.

B. Advanced Studies on Transition Zone

1. *Improvements for Electronic and Ionic Distributions*

Guidotti and Rice [18] have made further studies on the transition zone of Hg(liq.) by measuring the surface plasmon dispersion with an ATR (attenuated total reflection) technique [19-21] and proposed the model shown in Fig. 4c. According to this model (the GR-model), the optical property of the mercury surface is anisotropic. Namely, as we can see in Fig. 4c, the GR-model assumes that the optical conductivity parallel to the surface is large whereas the conductivity perpendicular to the surface is small in the transition zone. In addition, the GR-model adopts an oscillating electron density distribution $[\rho^-(z)/\rho_b^-]_{LK}$ proposed by Lang and Kohn [22]. Thus, the mathematical expression of the GR-model has been accomplished;

$$\sigma_x(z) = \sigma_y(z) = \sigma_b \left[\frac{\rho^-(z)}{\rho_b^-}\right]_{LK} \left[1 + A \exp\left\{-\frac{(z-z_0)^2}{\eta^2}\right\}\right]$$
$$\sigma_z(z) = \sigma_b \left[\frac{\rho^-(z)}{\rho_b^-}\right]_{LK} \tag{17}$$

where x and y represent two Cartesian coordinates parallel to the surface, while z represents the coordinate perpendicular to the surface, and A, z_0, and η are the characteristic parameters defining the Gaussian distribution function for the transition zone.

It is reported that a good agreement between the calculated optical property and the experimental result can be obtained by taking the following values for parameters involved in Eq. 17: $A = 6$, $z_0 = 2.31$ Å, $\eta = 0.3$ Å, and $(4\pi n/3)^{-1/3} = 5$, where n denotes the average number of electrons in a unit volume.

The incorporation of the oscillating electronic density distribution into the transition zone model is reasonable. As we can see in Fig. 4d, the density distribution of electrons confined in a volume should be oscillatory even for the metal with no special transition zone.

The distribution for the density of ions constituting the liquid metal is hitherto represented by a simple step function, as we can see in Fig. 4b-d (dotted line). This simplified model has to be revised, however, if two or three atomic layers at the surface of Hg(liq.) have special structures as suggested by the GR-model [18]. It must be pointed out that any ionic density distributions have to be consistent with the electron density distribution.

Theoretical calculation of the self-consistent density distributions of electrons and ions has been carried out for Hg(liq.) by Allen and Rice [23,24]. It has been shown that both ionic distribution and electronic distribution are oscillatory in the vicinity of the surface: The width of the transition zone has been estimated to be less than 7 Å. Although the results mentioned above are for the case of zero temperature, it has been suggested that the situations at high temperatures would qualitatively be the same as above. In connection with the above-mentioned discussion, it has also been suggested that, for a liquid metal with a small electron density, the shielding of ionic charge is insufficient and hence the interaction between ions and quasi-free electrons would be strong. Thus an oscillating electronic distribution can rather directly reflect upon the ionic distribution.

2. *Studies Based on X-Ray Reflection Data*

The special surface structures suggested by the theoretical analyses of optical properties of Hg(liq.) are stimulating for surface researchers and induce them to make more direct measurement for clarifying the surface structure of the liquid metal. The LEED (low energy electron diffraction) technique appears worth examining at first for this purpose because this technique might provide us with information about the distribution of atoms in the surface. Utilizing the LEED apparatus developed by themselves, Goodman and Somorjai [25,26] have attempted to take diffraction patterns of Pb(liq.), Bi(liq.), and Sn(liq.). Unfortunately, however, this has been unsuccessful.

On the other hand, the total reflection of X-ray has been shown to be applicable to the surface studies of liquid metals [27]. Since the refractive index of a liquid metal is less than 1 at the X-ray wavelength region, we can find a critical angle θ_c below which the total reflection of the X-ray takes place: The angle θ is defined as the angle incident to a smooth flat surface of a liquid metal. It is reported that, under ordinary conditions, θ_c takes a value of several milli-radian and X-rays with $\theta < \theta_c$ penetrate the liquid metal to a depth of ~10 Å from the surface [27,28].

Taking into account the characteristics of the X-ray reflection, Lu and Rice [27] have constructed a small-angle X-ray reflectometer. This apparatus is capable of resolving the reflection angle up to 10^{-4} radian and reaching a vacuum of $\sim 10^{-8}$ Torr: the Cr-Kα radiation (wavelength = 2.291 Å) has been used. With this apparatus, Lu and Rice have measured the X-ray reflection intensity for Hg(liq.) as a function of θ. The experimental results have revealed that the density profile in the transition zone of Hg(liq.) can be expressed by a hyperbolic tangent function of z. The width of the transition zone has been calculated to be 5.6 Å.

The transition zone model derived from the X-ray reflection data is nonoscillatory as mentioned above. On the contrary, D'Evelyn and Rice [29] have shown that the density profile in the transition zone of Hg(liq.) is oscillatory by a Monte Carlo simulation based on the pseudoatom theory. The simulation has suggested that (a) the density oscillation extends several layers into the bulk phase; (b) the oscillatory density profile has a spacing about one atomic diameter; (c) an adsorption layer exists in the metal-nonmetal transition zone.

The discrepancy between the result of Lu and Rice [27] and that of D'Evelyn and Rice [29] has been reconciled by the reinvestigation of the X-ray reflection data. It has been revealed that the angular dependence of the reflection intensity of the X-ray is rather insensitive to the surface structure, and hence both the oscillatory surface structure and the nonoscillatory surface structure can explain the experimental result [30]. In other words, the following oscillatory transition zone model (DR-model) can also explain the X-ray reflection data;

$$\frac{\rho(z)}{\rho_b} = \frac{1 + \left[\dfrac{-A_s \sin(2\pi z/\lambda_s)}{1 + (z/\ell_s)^2}\right]}{1 + \exp(-z/\delta)} + h_a \exp\left[\frac{-(z - z_a)^2}{w_a^2}\right] \quad (19)$$

where A_s, λ_s, ℓ_s, and δ are the parameters that determine the characteristics of the density oscillation; and h_a, z_a, and w_a are the parameters that characterize the Gaussian function representing the adsorption layer existing in the metal-nonmetal transition zone.

The density profile for the DR-model is illustrated in Fig. 4e. We can clearly see in this figure that an adsorption layer exists at $z = z_a$. It is reported that the appropriate values for the parameters are as follows: $A_s = 0.8$, $\lambda_s = 3.1$ Å, $\ell_s = 6.7$ Å, $\delta = 0.5$ Å, $h_a = 0.6$, $z_a = -3$ Å, and $w_a = 0.8$ Å. It must be pointed out, however, that another model with $A_s = 0 = h_a$ and $\delta = 1.42$ Å can also explain the experimental angular dependence of the X-ray reflection intensity [30], although Gorelsky et al [31] have reported that the angular dependence of the X-ray reflection coefficient is sensitive to the width of the electronic distribution profile at the surface.

As mentioned above, even with the X-ray reflection method it is hard to discriminate between the oscillatory and nonoscillatory transition zones. Thus, we need additional information to determine which one of the two models is appropriate.

Goodisman [32] has investigated this problem by calculating the work function of Hg(liq.). He has shown that a surface model similar to the DR-model gives a work function of 4.26 eV, close to the observed value of 4.5 eV. Monotonic profiles, including a step-function profile, have been shown to give much smaller values for the work function.

3. *Transition Zone for Other Liquid Metals*

Electronic and ionic density profiles for liquid metals other than Hg(liq.) have also been studied. The result for Al(liq.) [33,34] is shown in Fig. 4f. As we can see in the figure, both the ionic profile and the electronic profile are oscillatory and the largest amplitudes are found at the topmost layer of the surface. This result, however, does not provide us with information about the adsorption layer.

Monte Carlo simulations for the transition zones of Na(liq.) and Cs(liq.) have also been carried out [29,35,36]. It has been suggested that for each of these liquid metals there exists an oscillatory density distribution that extends several atomic layers from the surface to the interior of the bulk phase. The spacing of the oscillation is about one atomic diameter and the width of the transition zone is ~2.4-3.3 Å. A comparison of these simulations to that of Hg(liq.) has revealed that the ratio of the largest peak density to the bulk density at $z \to \infty$ is the largest for Hg(liq.), i.e., the ratios are ~3, ~1.5, and ~1.4 for Hg(liq.), Na(liq.), and Cs(liq.), respectively. From this result, a high regularity of the surface structure of Hg(liq.) has been suggested [30]. The surface structure of the Na-Cs liquid alloy has also been studied by the Monte Carlo simulation method [37]. According to the result of this simulation, the surface of the alloy is being covered by Cs atoms and the density profile is oscillatory.

In addition to the theoretical simulations mentioned above, experimental studies of the surface of Cs(liq.) have been carried out using the small-angle X-ray reflection method [38]. The previously constructed apparatus [27] has been improved to keep the Cs(liq.) surface clear: The base pressure of the sample chamber is 10^{-11} Torr (the working pressure is higher than this value because the vapor pressure of Cs is $\sim 10^{-6}$ Torr at 37°C). The experimental results have been consistent with the results of simulation mentioned above, i.e., the angular dependence of the X-ray reflection intensity can be explained by the oscillatory density profile with an adsorption peak outside the profile of the metallic phase.

C. Comments on the Transition Zone with Special Structure

Studies on the transition zone have originated from the experimental result that the ellipsometric optical properties for Hg-(liq.) are non-Drude-like, whereas reflectometric optical properties are Drude-like, as we have seen in the beginning of this section (Section IV). Inagaki et al [39] have criticized this and pointed out that the ellipsometric data are much more reliable than the reflectometric data. He has shown that no discrepancies appear between the ellipsometric results and the reflectometric results if the measurements are carefully carried out.

Another comment has been given by Greuter et al [40]. They have made a photoemission study of Hg(liq.) using an extremely pure sample (99.999995%) in an UHV cell: The temperature range is -84 to 22°C and the photon energy is 6.2 eV. The experimental result is the energy distribution curve (EDC) illustrated in Fig. 5. They have thus evaluated the ratio of the density of state at the Fermi level $N(E_F)$ to the area F under the EDC as a function of the angle of incidence. The resulting relation between $N(E_F)/F$ and the angle of incidence has been compared with the respective result of the theoretical calculation based on the ss-model, the BR-model, and the DR-model. It has been found that the ss-model gives the best agreement with the experimental

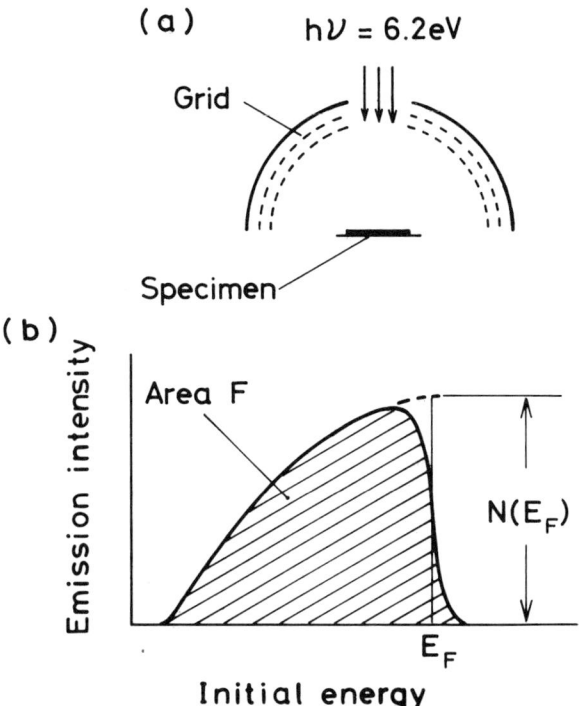

Fig. 5 (a) Principle of the photoemission study and (b) example of the energy distribution curve (EDC) [40].

result. It has been inferred, therefore, that the width of the transition zone is rather small, probably less than two atomic diameters.

REFERENCES

1. T. E. Faber, in *An Introduction to the Theory of Liquid Metals*, Cambridge University Press, 1972, p. 385.
2. J. N. Hodgson, in *Liquid Metals Chemistry and Physics* (S. Z. Beer, ed.), Marcel Dekker, Inc., New York, 1972, p. 331.
3. M. Shimoji, in *Liquid Metals*, Academic Press, London, New York, San Francisco, 1977, p. 333.
4. T. E. Faber and N. V. Smith, *J. Opt. Soc. Amer.*, 58, 102 (1968).

5. J. C. Miller, *Phil. Mag.*, *20*, 1115 (1969).
6. N. R. Comins, *Phil. Mag.*, *25*, 817 (1972).
7. E. D. Crozier and E. Murphy, *Can. J. Phys.*, *50*, 1914 (1972).
8. A. N. Bloch and S. A. Rice, *Phys. Rev.*, *185*, 933 (1969).
9. J. R. Beattie and G. K. T. Conn, *Phil. Mag.*, *46*, 222 (1955).
10. E. G. Wilson and S. A. Rice, *Phys. Rev.*, *145*, 55 (1966).
11. B. Siskind, J. Boiani, and S. A. Rice, *Can. J. Phys.*, *51*, 894 (1973).
12. L. G. Schulz, *J. Opt. Soc. Amer.*, *47*, 64 (1957).
13. J. N. Hodgson, *Phil. Mag.*, *4*, 183 (1959).
14. J. Boiani and S. A. Rice, *Phys. Rev.*, *185*, 931 (1969).
15. W. J. Choyke, S. H. Vosko, and T. W. O'Keeffe, *Solid St. Comm.*, *9*, 361 (1971).
16. J. N. Hodgson, *Phil. Mag.*, *7*, 229 (1962).
17. W. E. Mueller, *J. Opt. Soc. Amer.*, *59*, 1246 (1969).
18. D. Guidotti and S. A. Rice, *Phys. Rev.*, *B15*, 3796 (1977).
19. T. Lopez-Rios, *Opt. Comm.*, *17*, 342 (1976).
20. S. A. Rice, D. Guidotti, and H. L. Lemberg, *Adv. Chem. Phys.*, *27*, 543 (1974).
21. V. M. Agranovich and D. L. Mills, *Surface Polaritons*, North-Holland Pub. Co., Amsterdam, New York, Oxford, 1982.
22. N. D. Lang and W. Kohn, *Phys. Rev.*, *B1*, 4555 (1970).
23. J. W. Allen and S. A. Rice, *J. Chem. Phys.*, *67*, 5105 (1977).
24. J. W. Allen and S. A. Rice, *J. Chem. Phys.*, *68*, 5053 (1978).
25. R. M. Goodman and G. A. Somorjai, *J. Chem. Phys.*, *52*, 6325 (1970).
26. R. M. Goodman and G. A. Somorjai, *J. Chem. Phys.*, *52*, 6331 (1970).
27. B. C. Lu and S. A. Rice, *J. Chem. Phys.*, *68*, 5558 (1978).
28. D. Sluis and S. A. Rice, *J. Chem. Phys.*, *79*, 5658 (1983).
29. M. P. D'Evelyn and S. A. Rice, *Phys. Rev. Lett.*, *47*, 1844 (1981).
30. M. P. D'Evelyn and S. A. Rice, *J. Chem. Phys.*, *78*, 5081 (1983).
31. S. P. Goreslavsky, M. I. Ryazanov, R. C. Brown, and N. H. March, *Phys. Letts.*, *55A*, 123 (1975).

References

32. J. Goodisman, *J. Chem. Phys.*, *82*, 560 (1985).
33. O. A. Esin, V. A. Polukhin, and V. F. Ukhov, *Russ. J. Phys. Chem.*, *54*, 174 (1980).
34. L. V. Belan-Gaiko, V. I. Bogdanov, and D. L. Fuks, *Russ. J. Phys. Chem.*, *55*, 1048 (1981).
35. D. W. Oxtoby, F. Novak, and S. A. Rice, *J. Chem. Phys.*, *76*, 5278 (1982).
36. M. P. D'Evelyn and S. A. Rice, *J. Chem. Phys.*, *78*, 5225 (1983).
37. J. Gryko and S. A. Rice, *J. Phys. F: Metal Phys.*, *12*, L245 (1982).
38. D. Sluis, M. P. D'Evelyn, and S. A. Rice, *J. Chem. Phys.*, *78*, 1611 (1983).
39. T. Inagaki, E. T. Arakawa, and M. W. Williams, *Phys. Rev.*, *B23*, 5246 (1981).
40. F. Greuter, U. Gubler, J. Krieg, H. P. Preiswerk, and P. Oelhafen, *Surf. Sci.*, *124*, 489 (1983).

7
Electron Spectroscopies and Related Subjects

I.	Introduction	180
II.	Electronic Structures	181
	A. Photoemission Spectroscopy	181
	B. Electron Energy Loss Spectroscopy	187
	C. X-Ray Photoelectron Spectroscopy and Other Measurements	190
III.	Surface Composition	193
	References	199

I. INTRODUCTION

Applications of electron spectroscopic techniques to the studies of liquid metals and liquid alloys are expected to bring about fruitful results, as stressed by Evans [1]. The electron spectroscopic techniques coupled with ultrahigh vacuum (UHV) techniques enable us to monitor and to remove the surface contamination. Thus, we can obtain reliable information about the surface.

In the present chapter, electron spectroscopic studies of liquid metals and liquid alloys are reviewed first. It must be noted, however, that contents of this review do not always concentrate on the problems of surface only. In view of the importance of the problems involved, information about the bulk phase is also included. It is hoped, therefore, that readers will estimate the average depth of the source of information using the universal curve [2,3] representing the electron mean free path as a function of the electron kinetic energy, whenever necessary.

The second subject reviewed in this chapter is related to the analysis of the surface composition of the liquid alloy. Modern surface scientific techniques, including electron spectroscopies, are powerful tools for surface analysis and provide us with most reliable data regarding the surface composition. Although literature available for the present review has been limited, the usefulness of the application of new techniques would be understood.

II. ELECTRONIC STRUCTURES

A. Photoemission Spectroscopy

In the PES (photoemission spectroscopy), a light of a given energy is irradiated onto the specimen placed in a UHV chamber as shown in Fig. 1a, and the energies of electrons emitted from the specimen are analyzed. The number of photoelectrons measured as a function of the electron energy is usually expressed by a curve called EDC (energy distribution curve), which is illustrated in Fig. 1b. The EDC reflects the electronic density of states near the Fermi level when the energy of the incident light is low; it reflects the electronic structures of inner shells when the energy of the incident light is high.

The current situation of PES studies of liquid metals and liquid alloys can be seen in Table 1. The studies listed in this table have aimed mainly at clarifying whether the band structure of electron energies of a metal in the solid state collapses on melting. If the band structure is related to the long-range order of atomic arrangement in the solid state, it should collapse

TABLE 1 Photoemission Studies

Samples	Photon energy (eV)	Pressure[a] (Torr)	Temperature (°C)	Results	Ref.
In(liq.)	6.4-11.7	6×10^{-11}	185	Little differences between liquid and solid in density of states	4
In(sol.)	6.4-11.7	5×10^{-9}	Room temp.		
Au(liq.)	16.8, 21.2	2×10^{-8}	1125	Almost no change in the main peak position and in the peak width	5
Au(sol.)	26.9, 40.8	2×10^{-8}	25		
Hg(liq.)	5.5, 6.0	2×10^{-10}	(0)	Little differences in the main features of EDCs for liq. and sol.	6
Hg(sol.)	5.5, 6.0	2×10^{-10}	-113		
Cu(liq.)	16.6, 21.2	5×10^{-10}	1087	Little differences between EDCs for liq. and sol.[b]; a little difference between EDCs for liq. and sol.[c]	7
Cu(sol.)[b]	16.6, 21.2	5×10^{-10}	727		
Cu(sol.)[c]	16.6, 21.2	5×10^{-10}	--		
Ag(liq.)	21.2	1×10^{-10}	--	The 4d peak persists into the liquid state; EDC of liq. resembles that of sol.[d] but differs from that of sol.[e]	8
Ag(sol.)[d]	21.2	1×10^{-10}	--		
Ag(sol.)[e]	21.2	1×10^{-10}	--		
Hg(liq.)	10.2, 16.8, 21	4×10^{-9}	--	Some aspects of band structure in the solid state persist into liq.	9
In(liq.)	10.2, 21.2	10^{-9}	260	Little differences between EDCs for liquid and solid	10
In(sol.)	10.2, 21.2	10^{-9}	--		

Electronic Structures

Sample				Notes	Ref
Sn(liq.)	10.2, 21.2	10^{-9}	300	--	10
Hg(liq.)	9.33, 9.9, 10.2, 21.2	$2 \times 10^{4\,f}$	--	--	11
Al(liq.)	9.33, 9.9, 10.2, 21.2	$1 \times 10^{-9\,f}$	700	Imperfect agreement of EDC of Al(liq.) with that of Al(sol.), suggesting some aspects of solid electronic structure persist into liquid state	11
Ga(liq.)	21.2	5×10^{-10}	47-547	Little temperature dependence of spectra, being consistent with the similarity in the spectra above and below the melting point	12
Ga(liq.)	6.5	Better than 10^{-10}	-50 to -350	Liquid spectra are Drude-like while solid spectra are not, indicating an essential change of ionic short range order on melting	13
Ga(sol.)	6.5				

[a] Base pressure.
[b] Solid after frozen.
[c] Solid annealed.
[d] Solid after frozen and annealed.
[e] Evaporated film.
[f] Working pressure.

(a)

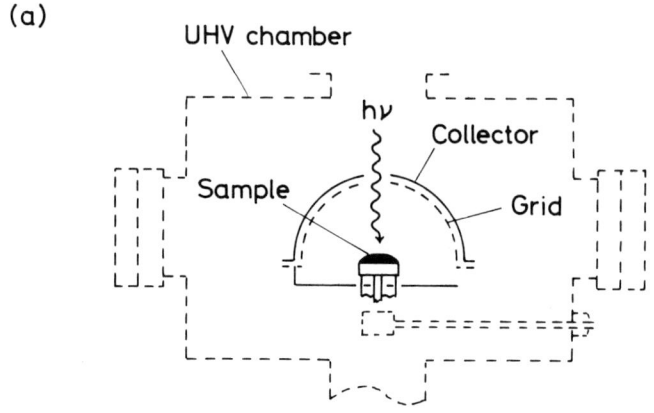

(b)

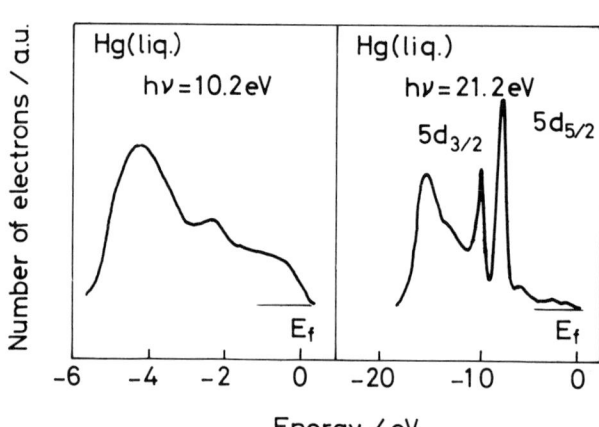

Fig. 1 An outline of the photoemission study of the liquid metal. (a) Photoemission cell placed in an UHV cell [10]; (b) EDCs for Hg(liq.) [11].

on melting. The experimental results are shown in Fig. 2a-c. This figure shows that the main feature of the EDC is preserved upon melting, although fine structures of the EDC vary on melting. This strongly suggests that the short-range ordering of atomic arrangement persists into the liquid state and governs the main feature of the EDC.

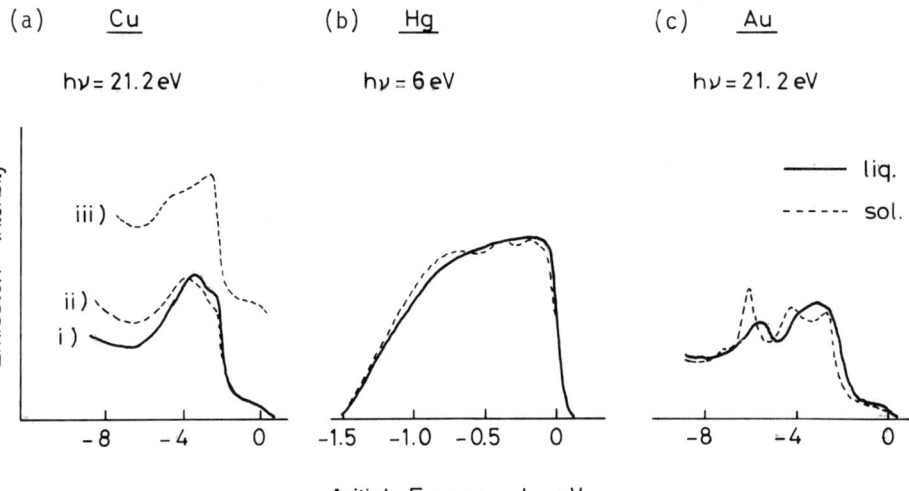

Fig. 2 EDCs for (a) Cu, (i) liquid (———) state at 1360 K, (ii) solid (----) state immediately after freezing at 1000 K, (iii) solid state after annealing at 1000 K [7]; (b) for Hg [6]; and (c) for Au [5].

Although most of the metals listed in Table 1 behave as described above, Al and Ga deserve special comment. The former metal, particularly in its molten state, does not give clear PE spectra because of its strong tendency to form an oxide [11]. This hinders us from knowing whether the electronic structure of Al persists into the liquid state. Fortunately, however, the result of a soft X-ray emission study of Al has proved that the short-range ordering or an instantaneous pseudocrystalline state persists into the liquid state [14]. On the other hand, the situation of Ga is different. It is reported that the structure of solid Ga at the melting point is quite special [13]. Namely, Ga at the melting point has an open structure with a low symmetry whereas liquid Ga has a normal, loose, close-packing structure. Thus, it is expected that the environment of each Ga atom would change drastically upon melting. In other words, a drastic change in the electronic structure of Ga is expected to occur on melting. Greuter and Oelhafen [13] have shown that the EDC near the Fermi level exhibits a remarkable change on melting.

The number of PES studies of liquid alloys is much less than that of the pure liquid metals. Only three reports were available for this review: one on the Ag-Cu system [8] and two on the In-Hg system [9,11]. The former study has been carried out mainly with a photon energy of 21.2 eV and with an UHV apparatus capable of keeping a 10^{-10} base pressure. It is reported that the Ag 4d peak in the EDC of the pure Ag reduces by 0.5 eV upon melting. This suggests that a reduction of the d-d overlapping has occurred on melting. Upon alloying Ag with Cu, the Ag 4d peak and the Cu 3d peak have shifted in opposite directions along the energy axis, suggesting s-charge transfers from Cu to Ag sites. For the In-Hg system, the alloying has resulted in broadenings and shifts of the EDC peaks for Hg $5d_{3/2}$ and Hg $5d_{5/2}$. The broadenings suggest that a reduction in the d-d overlapping has been brought about by the incorporation of In atoms into the Hg phase. The increase in the binding energy for the Hg 5d level (the EDC peak shift) suggests charge transfers from In to Hg sites [14]. It must be noted that the PES studies of liquid alloys enable us to know the surface composition [8]. As we can see in Section III of this chapter, a clear surface segregation of Ag has been observed for the liquid Ag-Cu alloy. This indicates that information about the surface is also involved in the PES data.

Another important quantity obtainable from the PES study is the work function of the liquid metal. It is interesting to see whether the work function varies on melting. The experimental results summarized in Table 2 clearly show that the work function varies little on melting. Norris [14] has postulated that this fact suggests a sharp and discontinuous density change at the surface of a crystalline metal persists into the liquid state.

TABLE 2   Work Function

Metals	Work function (eV) Liquid	Work function (eV) Solid	Temperature[a] (°C)	Ref.
In	3.94 ± 0.17	4.06 ± 0.10	--	14
In	4.14 ± 0.05	--	--	4
In	4.08 ± 0.04	4.08 ± 0.01	200	15
Hg	4.49 ± 0.01	4.49 ± 0.01	0	6
Al	4.02 ± 0.04	4.02 ± 0.04	--	14
Sn	4.27 ± 0.02	4.27 ± 0.02	--	14
Pb	3.94 ± 0.03	3.94 ± 0.03	--	14
Ga	4.35 ± 0.05	--	200	15
Ga	4.30 ± 0.02	--	50	13
Ga	$4.34 \pm 0.01 - (1.20 \pm 0.25) \times 10^{-4} T$	--	$300 \leq T \leq 623$ K	13
Bi	4.34 ± 0.05	--	300	15

[a]For liquid metals.

In the present stage, however, it is hard to make a rigorous theoretical calculation of work function of a polyvalent liquid metal [16].

B. Electron Energy Loss Spectroscopy

EELS (electron energy loss spectroscopy) has also been utilized in studying the electronic structure of the liquid metal [17-20]. In the EELS study, a primary electron beam with a given energy (E_p) is made incident to the surface of a specimen and the energies of reflected electrons are analyzed. A schematic drawing of the EELS measurement [18] and an example of the spectra [19] are shown in Fig. 3a and b, respectively. The spectra usually consist of peaks representing energy losses due to interband excitations of electrons [e.g., (i) and (ii) in Fig. 3b] and other peaks [e.g., (iii) in Fig. 3b] representing energy losses

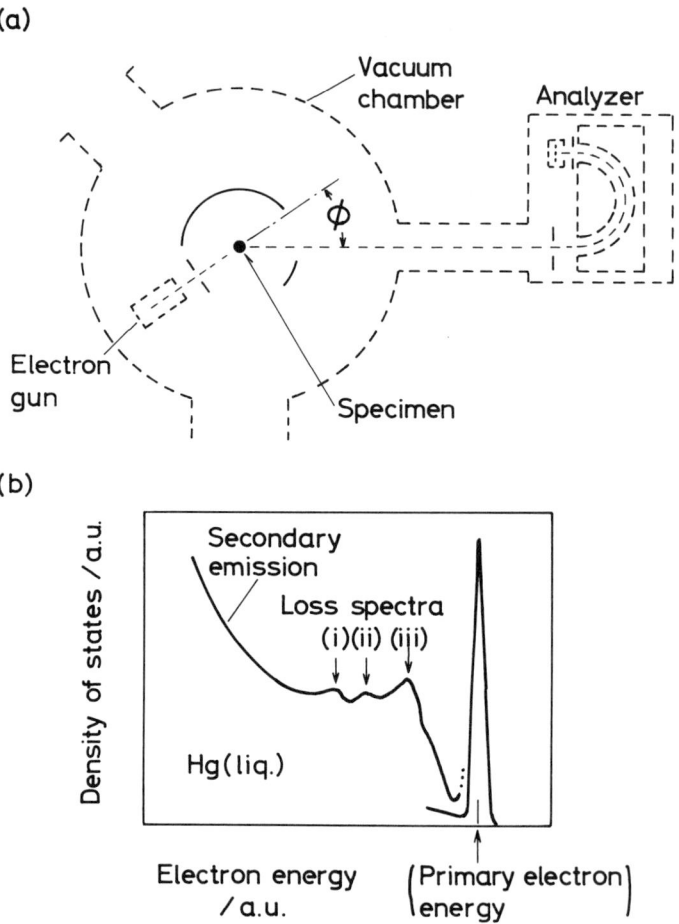

Fig. 3 An outline of the electron energy loss spectroscopic study of the liquid metal. (a) Schematic design of the apparatus [18]; (b) EEL spectra for Hg(liq.) [19].

due to collective motions of electrons in the specimen. The latter energy losses are called the plasmon losses and are classified as the volume plasmon loss and the surface plasmon loss. The volume plasmon loss spectra are seen when the angle of reflection φ is large (φ is defined in Fig. 3a). On the

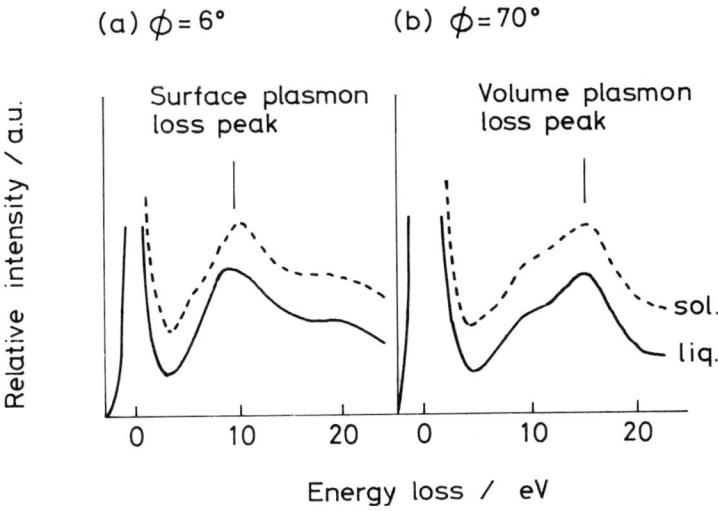

Fig. 4 Part of the EEL spectra of Bi: (a) surface plasmon loss peak; (b) volume plasmon loss peak [17].

other hand, the surface plasmon loss spectra are seen when φ is small. The angular dependences of the EEL spectra of Bi(liq.) [17] are shown in Fig. 4, to illustrate the situation mentioned above.

The main purposes of the EELS studies of liquid metals have been the investigation of the change in the electronic structure upon melting. With an expectation of a change in the electronic structure, Powell [17] has made a EELS study of Bi below and above the melting temperature using a primary electron energy of 8 KeV. It has been found that, upon melting, the surface plasmon loss peak at ~10 eV shifts toward lower energies while the volume plasmon loss peak at ~15 eV shifts toward higher energies. In addition, a peak at 5.3 eV disappeared and, instead, a peak appeared at 11.5 eV upon melting. From these facts, Powell has suggested that a change in the electronic structure of Bi might have occurred on melting. However, the change is inferred to be small because the main feature of the EEL spectra for Bi(liq.) differs little from the spectra for the solid Bi. Thus, he has extended the EELS studies using Al, In, Hg, Ga, Au, and Bi as samples [18]. The experimental results have shown that the main features of the EEL spectra for these metals change little on melting, suggesting that only small changes in the electronic structures have been brought about by the phase change from solid to liquid.

Guntherodt and his co-workers [19] have made EELS studies of Hg(liq.) and Ga(liq.) at E_p = 25 eV and E_p = 300 eV, respectively. For mercury, they found two interband excitation loss peaks, i.e., at 8.3 eV (Hg $5d_{5/2} \to E_F$) and at 10.4 eV (Hg $5d_{3/2} \to E_F$). In addition, a volume plasmon loss peak was found at 6.7 eV. It has been reported that the peak positions as well as the peak widths of above-mentioned spectra change little upon melting.

There are few EELS studies of liquid alloys in literature. Only Powell [20] has reported the results of EELS studies on the In-Al system and on the In-Bi system, both at 8 KeV. It was found that the electron energy losses due to the surface plasmon as well as those due to the volume plasmon take particular values depending upon the composition of alloy. According to the reports, the peak positions of the loss spectra do not agree with the values expected from the free electron theory.

In conclusion, the accumulation of EELS data appears insufficient for both liquid metals and liquid alloys. Studies under UHV conditions are particularly important for extracting meaningful information.

C. X-Ray Photoelectron Spectroscopy and Other Measurements

XPS (X-ray photoelectron spectroscopy) is a strong tool in surface studies and is widely used in the study of the heterogeneous catalysis. This situation is in sharp contrast to the difficulty in finding reports of XPS studies of liquid metals and liquid alloys. Namely, only one report has been available for this review.

Ichikawa [21] has made XPS studies of Au-Sn alloys of varying compositions (Mg-Kα excitation; temperature range, 300-350°C; vacuum level, 10^{-8} Torr). Measurements for the alloys in the solid states have also been carried out and the results have been compared with those for the liquid alloys. On the basis of the experimental results, Ichikawa discussed the electronic states of core electrons as well as those of the valence electrons.

Shown in Fig. 5 are the XP spectra for the core electrons, i.e., for the Sn 3d electrons and for the Au 4f electrons in the Sn-Au alloy containing 66.7 atom% of Sn. This figure serves for understanding the following description, which is the summary of the experimental results reported by Ichikawa [21].

1. The binding energy for the Sn 3d electrons decreases with the increase in the Sn content in the alloy.
2. At a given composition of the alloy, the binding energies for both Sn 3d and Sn 4d do not vary on melting.
3. The binding energy of the Au 4f core level increases with the increase in the Sn content of the alloy.
4. The binding energy of the Au 4f level for the solid state is somewhat larger than that for the liquid state, in the high Sn content region.
5. The energy separation between the Au $4f_{5/2}$ peak and the Au $4f_{3/2}$ peak does not change on melting.

The experimental results mentioned above show that the character of the core electrons changes little on alloying as well as on melting. According to Ichikawa, the changes in the binding energies with the change in the alloy composition (1 and 3), and the peak shift caused by the melting (4) as well, are

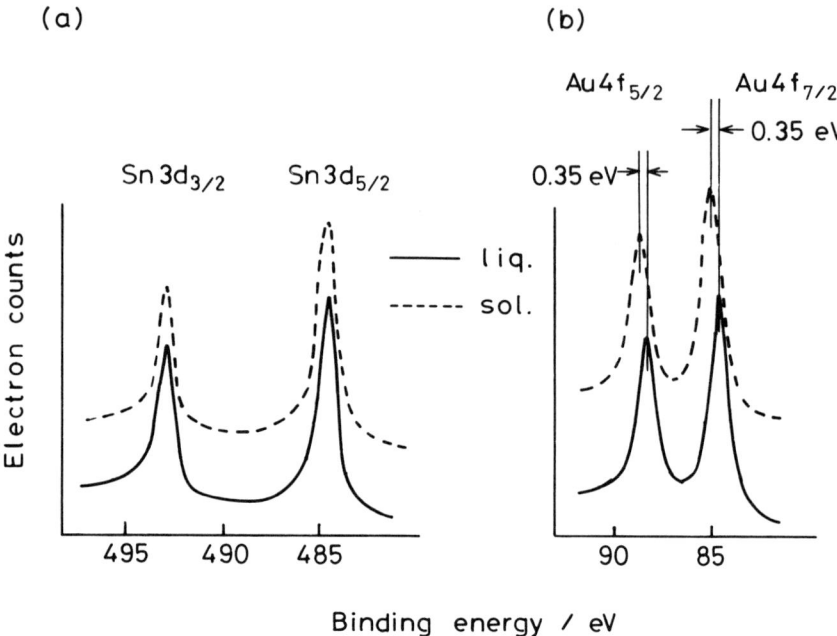

Fig. 5 XP spectra for the Sn-Au alloy containing 66.7 atom% of Sn: (a) Sn 3d doublet peaks; (b) Au 4f doublet peaks [21].

attributed to the change in the distribution of valence electrons. Both alloying and melting are expected to bring about a change in the atomic environment around a given element in the alloy. This would induce a change in the distribution of valence electrons and would result in a change in the potential acting upon the core electrons.

The XP spectra of the valence electrons for the Sn-Au alloy appear in the binding energy range of 0-10 eV: doublet peaks for Au 5d at 3-10 eV and a low plateau for Sn 5S and 5p at 0-3 eV. According to the report of Ichikawa [21], changes of these valence-band XP spectra upon melting and alloying have been consistent with the above-mentioned interpretation of the XP spectra for the core electrons.

Apart from the XPS studies, there are a few optical studies that aim at clarifying electronic structures of liquid alloys. Guntherodt and his co-workers [19] have studied the electronic structure (valence band) of the Au-Sn alloy by means of an optical reflection. According to their results, the Au 5d band width becomes broader and its center of gravity shifts toward lower energies as the Au content of the liquid alloy increases. These results agree partly with the results obtained by the XPS study of Ichikawa [21]. Krohn and Thompson [22] and Fainchtein et al [23] have published results of optical studies on several liquid alloys: Ga-Te, Tl-Te, In-Te [22]; Cs-Tl [23]. The purposes of these reflectivity measurements (normal incidence) are to obtain information available for making energy band models. The band models reported by these authors are useful for discussing the metal-semiconductor transition in the liquid state.

III. SURFACE COMPOSITION

The determination of the surface composition is indispensable for making a quantitative discussion of any surface phenomenon. Thus, a number of surface analytical techniques have been explored and are being utilized in many research fields related to the solid surfaces [24-28]. In contrast to this, the modern surface analytical techniques have only partly been applied to the surface analyses of liquid alloys, as we can see in Table 3. Part of this situation can be attributed to the drawbacks of studying the liquid metals or liquid alloys in the UHV apparatus. For instance, the high temperature necessary for keeping the specimen in a liquid state together with the vapor of the metallic element would make the maintenance of a high-quality UHV apparatus difficult. In addition, we have to place the specimen on a horizontal specimen holder because the specimen is in a liquid state. This sometimes requires us to use a specially designed apparatus [33,34].

TABLE 3 Studies on Surface Composition

Liquid alloys	Methods	Signal; Peak position	(eV)	Contamination	Cleaning by	Pressure (Torr)	ss+	Ref.
In-Pb (4.55-60.49)[a]	AES	In: $M_5N_{4,5}N_{4,5}$; Pb: NOO;	403 92	C, S	Ion bombardment Oxidation of C by O_2	10^{-9}	Pb	29
Al-Sn (0.62-10.53)[b]	AES	Al: $L_{2,3}VV$; Sn: $M_5N_{4,5}N_{4,5}$;	68 430	C, O, S	Flash heating at 873 K and Ar^+ bombardment	Better than 10^{-7}	Sn	30
Al-Cu (19.7-60.9)[b]	AES	Al: $L_{2,3}VV$; Cu: $L_3M_{4,5}M_{4,5}$;	68 920	C, O, S	Same as above	As above	Al	30, 31
Ga-In (16.5)[a]	AES	Ga: $L_3M_{4,5}M_{4,5}$; In: $M_5N_{4,5}N_{4,5}$;	1069 403	Oxide of Ga	Ar^+ bombardment	Better than 10^{-8}	In	32
Ga-Sn (0-0.7)[c]	AES	Ga: -- Sn: --	1069 430	O, C	Ar^+ bombardment	--	Sn	33
In-Sn	AES	In: $M_5N_{4,5}N_{4,5}$; Sn: $M_5N_{4,5}N_{4,5}$;	403 430	O, C	Ar^+ + H_2 bombardment	10^{-8}	In	34
Ga-Sn (5.13-83.25)[a]	AES	--	--	--	--	--	Sn	35

Surface Composition

Binary alloys; combination of In, Ga, Sn, Pb, and Bi	Technique	Lines	Contamination	Treatment	Working pressure	Segregated	Atom%	Ref
Ag-Cu	AES	--	O, C	--	--	--	--	36
	UPS (He I)	Ag: 4d; Cu: 3d;	O	Heating in H_2 or in vacuum	10^{-10} ($\sim 10^{-8}$)[d]	Ag	8	
Au-Sn	XPS (Mg-Kα)	Au: 4f; Sn: 3d;	C, Cl	Ar^+ bombardment	2×10^{-8}[e] 3×10^{-9}[f]	Sn	21	
Ga-In (Eutectic)	ISS (2 KeV Ar^+)	--	Oxide of Ga	Ar^+ bombardment	3×10^{-8}	In	32	

[†] Surface segregation.
[a] Atom% of the second element.
[b] Weight% of the second element.
[c] Atomic fraction of the second element.
[d] Working pressure.
[e] 300–350°C.
[f] 50°C.

AES (Auger electron spectroscopy) has most frequently been applied to the surface analysis of the liquid alloy, as we can see in Table 3. In the AES analytical studies, many authors have pointed out that eliminations of contaminants from the surfaces are not easy. The most common contamination elements are carbon and oxygen, as revealed by the data shown in Table 3. Usually the contamination is removed by an ion bombardment. This cleaning technique, however, has to be applied carefully, in particular to the alloy containing elements of different sputtering yields: A long period of sputtering would result in a change in the composition of the specimen. It must be noted, in addition, that the ion bombardment is sometimes insufficient for a complete elimination of the surface contamination [29,30,34]. In such cases, we have to apply some special methods to remove the contamination. For instance, Berglund and Somorjai [29] have adopted an oxidation method to remove carbon from the surface of the In-Pb liquid alloy. It must be noted, however, that this method cannot be used at temperatures higher than 873 K where the vaporization of Pb is significant. In addition, the oxidation method is not effective for the alloys of higher Pb contents because the formation of PbO competes with the carbon elimination. Another method of removing the surface contamination has been reported by Tsukamoto [34]. According to his report, it is difficult to remove oxygen from the surface of the In-Sn liquid alloy by an argon ion bombardment. The argon ion bombardment in the presence of hydrogen has been found to be effective to remove oxygen (see Fig. 24, Chapt. 4).

The surface composition of a liquid alloy can be determined from the experimentally obtained intensity of the AE spectral peak. For instance, Goumiri [30] has determined the surface composition of the Sn-Al alloy using the following relation:

$$I_{Sn} = I_{Sn}^{\infty} X_{Sn}^{s} \left(\frac{1 + r_{Al}}{1 + r_{Sn}} \right) \left[1 - \exp\left(-\frac{m}{0.74 \lambda_{Sn}} \right) \right] \tag{1}$$

where I_{Sn} is the Auger peak intensity observed for the liquid alloy, I_{Sn}^{∞} is the Auger peak intensity for pure Sn(liq.), X_{Sn}^{s} is the atomic fraction of Sn in the surface phase, $1 + r_j$ is the factor representing the back-scattering of Auger electrons for the j-component, λ_j is the inelastic mean free path of Auger electrons for the j-component, m is the number of atomic layers constituting the surface phase, and the numerical factor 0.74 is the constant determined from the geometry of the apparatus (the axis of CMA vertically crosses the surface of the specimen) [33]. It must be noted that we can approximate the value for I_j^{∞} by the Auger signal peak for the pure j-component in the solid state when the melting point of j is high, otherwise a high vapor pressure of j in the molten state hinders us from taking the AES data for the pure j-component in the liquid state. Another point to be noted here is that the inclusion of the factor m into Eq. 1 expresses the possibility that the Auger electrons come not only from the topmost layer of the specimen but also from several atomic layers below the surface. According to the experimental results, however, the surface composition determined by the AES method agrees fairly well with those determined by other methods (e.g., the thermodynamic or statistical thermodynamic calculation based on the surface tension data), when we assume that m = 1 [30-32].

Although the primary purpose of either the UPS (ultraviolet photoelectron spectroscopy) study or the XPS study is to investigate the electronic structure of the specimen, the experimental data can serve for calculating the surface composition. This is of great significance because we need exact information about the surface composition in order to discuss the electronic structure on the basis of UP or XP spectral data.

The UPS study of the Ag-Cu liquid alloy reported by Williams and Norris [8] exemplifies the use of the spectral data for the surface analysis. By assuming equal cross sections, they have

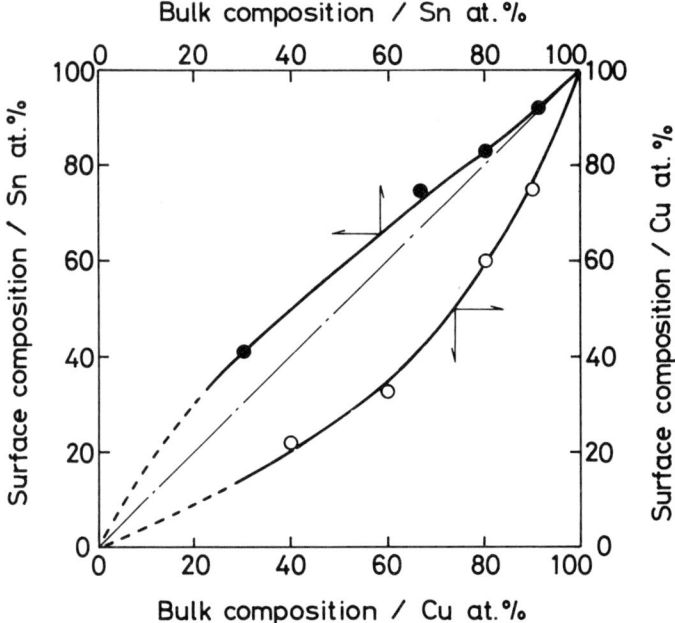

Fig. 6 Two examples of the surface enrichment in the binary liquid alloy: (o) Ag-Cu system [8]; (●) Au-Sn system [21].

plotted the Cu concentration derived from the relative intensity of the Cu 3d peak against the Ag 4d peak as a function of the bulk concentration of Cu. The resulting curve has been concave, as we can see in Fig. 6 (the calculated Cu concentration is defined as the surface concentration and is given by the ordinate of this figure). A comparison of this curve with the result of thermodynamic calculation has proved that the concave relation mentioned above is due to the surface enrichment of Ag.

The work of Ichikawa [21] exemplifies the use of the XPS data for the surface analysis of the liquid alloy. From such XP spectra, as shown in Fig. 5, the respective spectral intensities for Sn (I_{Sn}, the sum of the integral intensities of Sn $3d_{3/2}$ and Sn $3d_{5/2}$ peaks) and for Au (I_{Au}, the sum of the integral intensities of Au $4f_{5/2}$ and Au $4f_{7/2}$ peaks) have been obtained.

Ichikawa has determined the surface composition of the Sn-Au liquid alloy by putting the XP spectral intensities I_{Sn} and I_{Au} into the following relation:

$$X_{Sn}^S = \frac{I_{Sn}}{I_{Sn} + kI_{Au}} \quad (2)$$

where k is the ratio of the XP spectral intensity of pure Sn to that of pure Au.

As we can see in Fig. 6, the surface of the Sn-Au liquid alloy is somewhat enriched by the Sn component. It must be noted, however, that the 'surface' does not always mean the topmost layer of the liquid alloy. According to Ichikawa, the probe depth for the XP process reaches several tens of angstroms.

For the purpose of analyzing the topmost layer of the specimen, the use of ISS (ion scattering spectroscopy) with a noble gas ion having a few KeV, or less, of energy appears appropriate [32] because ions penetrating to the second, or more atomic layers, from the topmost layer are neutralized. Dumke [32] has thus applied the ISS technique to the measurement of the surface composition of the eutectic alloy of In-Ga. The experimental result revealed that the surface concentration of Ga for the In-Ga liquid alloy is less than 6%, indicating a strong surface enrichment of In (the bulk concentration of Ga is 83.5%). Dumke et al have compared this result with the result of a scanning Auger spectroscopy. According to the Auger data, the best estimate of the surface concentration of In for the eutectic In-Ga liquid alloy was 98%, in good agreement with the ISS data.

REFERENCES

1. R. Evans, *J. Physiq.*, 41, C8-775 (1980).
2. G. A. Somorjai, *Chemistry in Two Dimensions: Surfaces*, Cornell University Press, Ithaca and London, 1981, p. 41.

3. M. P. Seah, in *Practical Surface Analysis by Auger and X-Ray Photoelectron Spectroscopy* (D. Briggs and M. P. Seah, eds.), John Wiley & Sons, Chichester, New York, Brisbane, Toronto, Singapore, 1983, p. 186.
4. R. Y. Koyama and W. E. Spicer, *Phys. Rev. B*, *4*, 4318 (1971).
5. D. E. Eastman, *Phys. Rev. Letts.*, *26*, 1108 (1971).
6. P. Cotti, H.-J. Guntherodt, P. Munz, P. Oelhafen, and J. Wullschleger, *Solid St. Comm.*, *12*, 635 (1973).
7. G. P. Williams and C. Norris, *J. Phys. F: Metal Phys.*, *4*, L175 (1974).
8. G. P. Williams and C. Norris, *Phil. Mag.*, *34*, 851 (1976).
9. C. Norris, D. C. Rodway, G. P. Williams, and J. E. Enderby, *J. Phys. F: Metal Phys.*, *3*, L182 (1973).
10. C. Norris, D. C. Rodway, and G. P. Williams, in *The Properties of Liquid Metals* (S. Takeuchi, ed.), Taylor and Francis Ltd., London, 1973, p. 181.
11. C. Norris, D. C. Rodway, and G. P. Williams, *J. Physiq.*, *35*, C4-61 (1974).
12. C. Norris and J. T. M. Wotherspoon, *J. Phys. F: Metal Phys.*, *7*, 1599 (1977).
13. F. Greuter and P. Oelhafen, *Z. Physik.*, *B34*, 123 (1979).
14. C. Norris, *Inst. Phys. Conf. Ser.*, No. 30, 171 (1977).
15. K. B. Khokonov, S. N. Zadumkin, and B. B. Alchagirov, *Elektrokhimiya*, *10*, 911 (1974).
16. T. E. Faber, *An Introduction to the Theory of Liquid Metals*, Cambridge University Press, 1972, pp. 401-404.
17. C. J. Powell, *Phys. Rev. Letts.*, *15*, 852 (1965).
18. C. J. Powell, *Phys. Rev.*, *175*, 972 (1968).
19. H. J. Guntherodt, P. Oelhafen, R. Lappka, H. U. Kunzi, G. Indlekofer, J. Krieg, T. Laubscher, H. Rudin, U. Gubler, E. Hauser, M. Liard, M. Muller, J. Kubler, K. H. Bennemann, and C. F. Haugue, *J. Physiq.*, *41*, C8-381 (1980).
20. C. J. Powell, *Adv. Phys.*, *16*, 203 (1967).
21. T. Ichikawa, *Phys. Stat. Sol.*, *(a)32*, 369 (1975).
22. C. E. Krohn and J. C. Thompson, *J. Phys. F: Metal Phys.*, *11*, 1811 (1981).
23. R. Fainchtein, U. Even, C. E. Krohn, and J. C. Thompson, *J. Phys. F: Metal Phys.*, *12*, 633 (1982).

24. T. A. Carlson, *Photoelectron and Auger Spectroscopy*, Plenum Press, New York and London, 1975.
25. T. M. Buck, in *Methods of Surface Analysis* (A. W. Czanderna, ed.), Elsevier Scientific Publishing Company, Amsterdam, Oxford, New York, 1975.
26. R. Vaselow and S. Y. Tong, *Chemistry and Physics of Solid Surfaces*, CRC Press, Inc., USA, 1977.
27. M. Thompson, M. D. Baker, A. Christie, and I. F. Tyson, *Auger Electron Spectroscopy*, John Wiley & Sons, New York, Chichester, Brisbane, Toronto, Singapore, 1985.
28. R. K. Grasselli and J. F. Bazdil, *Solid State Chemistry in Catalysis*, Amer. Chem. Soc., Washington, D.C., 1985.
29. S. Berglund and G. A. Somorjai, *J. Chem. Phys.*, *59*, 5537 (1973).
30. L. Goumiri, P. Laty, J. C. Joud, and P. Desre, *J. Physiq.*, C8-787 (1980).
31. P. Laty, J. C. Joud, and P. Desre, *Surf. Sci.*, *104*, 105 (1981).
32. M. F. Dumke, T. A. Tombrello, R. A. Weller, R. H. Housley, and E. H. Cirlin, *Surf. Sci.*, *124*, 407 (1983).
33. S. Hardy and J. Fine, *Surf. Sci.*, *134*, 184 (1983).
34. H. Tsukamoto, Ms. thesis, Tohoku University, 1985.
35. A. A. Shebzukhov and O. A. Ashkhotov, *Poverkh, Fiz. Khim.*, Mekh, 64 (1983).
36. O. A. Ashkhotov, A. A. Shebzukhov, and K. B. Khokonov, *Dokl. Akad. Nauk. SSSR*, *274*, 1349 (1984).

Author Index

Numbers in parentheses are reference numbers and indicate that an author's work is referred to although the author's name is not cited in the text. Underlined numbers give the page on which the complete reference is listed.

A

Abbachian, G. J., 127(40), 130(40), 152
Abraham, F. F., 126, 151
Aczel, T., 60(47), 71
Adachi, A., 136(57), 153
Adadurow, I. E., 2(5), 7
Addison, C. C., 23, 23, 63, 64(70,76), 71, 72, 111, 124
Addison, W. E., 63(53-55), 71
Agranovich, V. M., 171(21), 178
Alchagirov, B. B., 115(49), 124, 127(37,39), 152, 187(15), 200
Allen, B. C., 126, 127, 151
Allen, J. W., 172, 178
Ambwami, D. S., 145(83,84), 154
Arakawa, E. T., 176(39), 179
Ashcroft, N. W., 127(30), 152
Ashe, T. R., 60(44), 71
Ashkhotov, O. A., 194(35,36), 201
Averbach, B. L., 96(28), 97(28), 123

B

Baker, M. D., 193(27), 201
Baker, M. G., 63(62), 71
Balooch, M., 66, 72, 111, 124
Barton, B. D., 68(82), 72
Baryshev, E. E., 141(70), 153
Baum, B. A., 141(70), 153
Bazdil, J. F., 193(28), 201
Bazin, Y. A., 141(70), 142(72), 154
Beattie, J. R., 160(9), 178
Beer, S. Z., 125(2), 150
Belan-Gaiko (Belan-Gayko), L. V., 127, 151, 175(34), 179
Belov, B. T., 130(47), 152
Belton, G. R., 144, 154
Bennemann, K. H., 187(19), 190(19), 193(19), 200
Berglund, S., 194(29), 196, 201
Bertrand, P. A., 141(69), 142(69), 153
Bhargava, B. S., 134(54), 153
Blairs, S., 127(35), 130(35), 148, 152, 154
Bloch, A. N., 160(8), 161, 163(8), 168, 178

203

Bogdanov, V. I., 127(15), 151, 175(34), 179
Boiani, J., 161(11), 163(11, 14), 170(11), 178
Bolshov, A. L., 142(75), 154
Boom, R., 148, 150(86), 154
Brown, J. K., 52(35), 70
Brown, O. W., 2(3), 6
Brown, R. C., 131, 152, 174(31), 178
Buck, T. M., 193(25), 201

C

Caracciolo, G., 127(42), 152
Carnahan, W. F., 133(53), 153
Carson, T. A., 193(24), 201
Chacon, E., 127, 151, 152
Chegodayev, A. I., 134(63), 141(63), 153
Chentsov, V. P., 134(62), 153
Chernov, A. I., 135(66), 153
Choyke, W. J., 164(15), 178
Christie, A., 193(27), 201
Chumarev, V. M., 134(62), 153
Cirlin, E. H., 197(32), 199(32), 201
Cochran, S. J., 55(37), 56, 70
Coldrey, J. M., 63(57,58), 71
Comins, N. R., 158(6), 159(6), 178
Conn, G. K. T., 160(9), 178
Cordovilla, C. G., 130(48), 131(48), 152
Cotti, P., 182(6), 200
Couchman, P. R., 142(71), 153
Creffield, G. K., 63(63,67), 72
Cromer, D. T., 97(31), 123
Cronauer, D. C., 83(16), 123
Croxton, C. A., 126, 133(12), 151
Crozier, E. D., 159(7), 160(7), 161(7), 163, 170, 178
Cusumano, J. A., 49(33), 70

D

Dalla Betta, R. A., 49(33), 70
Davies, B. M., 63(64-66), 72
Demeri, M., 134(60), 142(60), 153
Desre, P., 127(36,38), 134(38), 137(59), 142(74,77), 152-154, 194(30,31), 196(30), 197(30,31), 201
D'Evelyn, M. P., 173, 174, 175(29,30,36), 178, 176(38), 179
Didenko, P. D., 2, 7
Domashnikov, B. P., 141(70), 142(72), 153, 154
Donaldson, M., 60(44), 71
Dranchuk, M. M., 130(46), 152
Dubinin, E. L., 134(63), 141(63), 153
Dumke, M. F., 194(32), 197(32), 199, 201
Dyer, C. W., 60(46), 71

E

Eastman, D. E., 182(5), 200
Egelstaff, P., 96(29), 123
Elkin, E. M., 2(4), 6
Ellus, D., 62(49), 71
Enderby, J. E., 96(29), 123, 182(9), 200
Esin, O. A., 175(33), 179
Eustathopoulos, N., 142(77), 154
Evans, R., 126, 127(17,18,21), 151, 152, 180, 199
Even, U., 193(23), 200
Ewing, C. T., 81(13), 122
Eyring, H., 77(5), 122

F

Faber, T. E., 115(50), 124, 125(1), 126, 150, 155(1), 158, 177, 187(16), 200

Fainchtein, R., 193, 200
Falke, W. L., 136(58), 153
Farag, M., 134(60), 142(60), 153
Farcasiu, M., 58(38), 70
Ferrier, R. P., 126, 151
Fine, J., 193(33), 197(33), 201
Flores, F., 127(23,25,32), 151, 152
Foiles, S. M., 127(30), 152
Fort, T., Jr., 145(83,84), 154
Fowler, R. H., 126, 151
Friedman, S., 3(7), 7, 49(31), 65, 70, 72
Fuks, D. L., 127(15), 151, 175(34), 179

G

Galloisand, B., 135(65), 153
Gidaspow, D., 12, 23
Glasstone, S., 77(5), 122
Goodisman, J., 127(26), 151, 175, 179
Goodman, R. M., 173, 178
Goreslavsky, S. P., 174(31), 178
Goumiri, L., 127(38), 134(38), 142(74,76), 152, 154, 194(30), 196, 197(30), 201
Grasselli, R. K., 193(28), 201
Greuter, F., 176, 179, 183(13), 185(13), 186, 200
Grimson, M. J., 127(29), 152
Gryko, J., 175(37), 179
Gubler, U., 176(40), 179, 187(19), 190(19), 193(19), 200
Guidotti, D., 168(18), 171, 172(18), 178
Guntherodt, H.-J., 182(6), 187(19), 190, 193, 200

H

Halder, N. C., 97(32), 123
Halstead, W. D., 63(57), 71
Hamamatsu, S., 136(57), 153
Hardy, S., 193(33), 197(33), 201
Harrison, D. A., 127(35), 130(35), 148, 152
Hartman, R. J., 2, 6
Hasegqwa, M., 127, 127(20,21,24), 151
Hatswell, M., 55(37), 56(37), 70
Hauge, C. F., 187(19), 190(19), 193(19), 200
Hauser, E., 187(19), 190(19), 193(19), 200
Heasley, J., 134(60), 142(60), 153
Hicter, J. M., 127(38), 134(38), 152
Hiramatsu, N., 10(5), 23, 25(4), 31(4), 69
Hoar, T. P., 106, 124
Hobdel, M. R., 63(69-71,74), 64(70), 72
Hodgson, J. N., 155(2), 161(13), 163(13), 166(16), 177, 178
Holbrook, K. A., 58(41), 71, 79(9), 122
Honda, K., 26(19,20,21), 37(19,20,21), 43(21), 69, 70, 112(48), 124
Housley, R. H., 194(32), 197(32), 199(32), 201
Hubberstey, P. H., 63(63,67, 72,73), 72
Hultgren, R., 103(37), 106, 124

I

Iberson, E., 63(56), 71
Ibragimov, K. I., 134, 153
Ichikawa, T., 191-193, 198, 200
Iida, T., 127(33), 149, 152, 154
Inagaki, T., 176, 179
Indlekoffer, G., 187(19), 190(19), 193(19), 200

Ipatiew, W., 2, 6
Isherwood, S. P., 101(34), 123
Ishikawa, K., 26(27), 56(27), 70, 87(19), 123
Itami, T., 133, 153

J

Jackson, W. R., 55(37), 56(37), 70
Jacob, K. T., 131, 153
Jao, R. A., 145(84), 154
Josi, M. L., 96(26), 123
Joud, J. C., 127(36,38), 134(38), 137(59), 142(74,76,77), 152-154, 194(30), 196(30), 197(30,31), 201

K

Kamiya, Y., 58(39,40), 71
Kaplow, R., 96(28), 97(28), 123
Kasama, A., 127(33,43), 130(49), 133(49), 143, 136(57), 149(89), 152-154
Kashiwadate, K., 10(4), 23, 25(3), 30(3), 69
Kaufman, M. L., 65(77), 72
Kawai, Y., 144(79), 154
Kawanami, N., 10(5), 23, 25(4), 31(4), 69
Kelley, D. F., 68(82,83), 72, 73
Kemball, C., 77, 78, 122, 145, 154
Kerridge, D. H., 63(52-55), 71
Khanna, K. N., 134(54), 153
Khokonov, K. B., 115(49), 124, 127(37,39), 152, 187(15), 194(36), 200, 201
Kishimoto, M., 144, 154

Kisil, I. S., 130(46), 152
Kita, Y., 136(47), 153
Kohn, W., 171, 178
Kokov, M. B., 127(37), 152
Komiyama, M., 26(25,28,29), 49(25,34), 53(34), 58(34), 62(28,29), 70
Kondo, S., 126, 126(5), 151
Konovalov, V. A., 141(70), 142(72), 153, 154
Kovalchuk, V. F., 106, 124
Koyama, R. Y., 182(4), 200
Krieg, J., 176(40), 179, 187(19), 190(19), 193(19), 200
Krohn, C. E., 193, 200
Kubler, J., 187(19), 190(19), 193(19), 200
Kumaravadivel, R., 127(17,18), 151
Kunzi, H. U., 187(19), 190(19), 193(19), 200

L

Ladner, W. R., 52(35), 70
Laidler, K. J., 77(5), 122
Lang, G., 127(36,41), 132(55), 137(59), 148, 149, 152-154
Lang, N. D., 171, 178
Lange, K. W., 127(44), 152
Lappka, R., 187(19), 190(19), 193(19), 200
Larkins, F. P., 55(37), 56(37), 70
Laty, P., 127(36), 137(59), 141(59), 152, 153, 194(30), 196(30), 197(30,31), 201
Laubscher, T., 187(19), 190(19), 193(19), 200
Lemberg, H. L., 171(20), 178
Lepinskikh, B. M., 127, 151
Levy, R. B., 49(33), 70
Lewis, J., 63(52,53), 71
Liard, M., 187(19), 190(19), 193(19), 200

Lieberman, H. H., 135(67), 153
Lindemann, F. A., 149, 154
Lopez-Rios, T., 171(19), 178
Louis, E., 130(48), 131(48), 152
Lu, B. C., 173, 174, 176(27), 178
Lupis, C. H. P., 135(65), 153

M

McLean, A., 127(34,43), 130(34), 143(43), 152
Makino, K., 49(32), 70
Malko, A. G., 130(46), 152
Manning, J. A., 63(56), 71
March, N. H., 131, 152
Marley, N. F., 66(78), 72
Marquardt, D., 85, 123
Martin, H. H., 3(6), 7, 25(9), 69
Maslennikov, Y. I., 96(24), 123
Masuda, T., 115(51), 124
Matsunaga, S., 20(12), 23, 26(22-24), 49(23,34), 51(23), 53(34,36), 58(34), 70, 83(17), 85(17), 123
Matsuura, M., 20(12), 23, 26(22-24), 49(23), 51(23), 70, 83(17), 85(17), 123
Meszaros, L., 4, 7
Michell, T. O., 58(38), 70
Miedema, A. R., 150(86), 154
Migai, L. L., 135(66), 153
Mikhailov, N. Y., 135(66), 153
Miller, J. C., 158(5), 159(5), 178
Miller, R. R., 81(13), 122
Miller, V. R., 136(58), 153
Miller, W. A., 127(34,43), 130(34), 143(43), 152
Mills, D. L., 171(21), 178
Misawa, M., 149(89), 154

Mitko, M. M., 134(63), 141(63), 153
Mittag, U., 127(44), 152
Miyamoto, A., 10(3,4), 12, 14, 16(10), 17(10), 23, 25(2,3,5,12,13), 26(6), 30(2,3), 47, 49(13), 69, 75(1,2), 78(1,2,7,8), 79(2,8), 87, 105(38), 106(38), 112(46), 115, 118(2,46), 119(2,46), 122-124
Miyashita, F., 9(1), 15(1), 23, 25(6), 26(6), 69, 75(3), 78(3), 122
Mon, K. K., 127(19), 151
Mori, K., 144(79), 154
Morita, Z., 127(33,43), 130(49), 133(49), 136(57), 143(43), 149(89), 152-154
Mueller, W. E., 166(17), 178
Muller, M., 187(19), 190(19), 193(19), 200
Munz, P., 182(6), 200
Murphy, E., 159(7), 160(7), 161(7), 163, 170, 178

N

Nason, D., 142(73), 154
Nasyyrov, Y. A., 142(72), 154
Navascues, G., 127(23,25,32), 151, 152
Nilsson, W. B., 68, 72
Nogi, K., 143(78), 154
Nojima, S., 26(28,29), 62(28,29), 70
Nordine, P. C., 66(78), 72
Norris, C., 112, 115(47), 124, 182(7-10), 183(10-12), 185(11,14), 186, 197, 200
North, D. M., 96(29), 123
Novak, F., 175(35), 179

O

Oberburg, S. H., 141(69), 142(69), 153
Oelhafen, P., 176(40), 179, 182(6), 183(13), 185(13), 186, 187(19), 190(19), 193(19), 200
Ogino, K., 143(78), 154
Ogino, Y., 6(12), 7, 9(1,2), 10(3-6), 12, 14, 15(1), 16(10), 17(10,11), 20(12), 23, 25(2-8, 11-13,15,16), 26(5,6, 17-30), 30(2,3), 31(4), 33(7,8), 35(7,8), 36(8, 17,18), 37(19-21), 43(21), 46(11,16), 47, 49(13,25,34), 51(23), 53(34), 56(27), 57(26), 58(34), 62(28-30,51), 69-71, 75(1,3), 78(1,3, 7,8), 79, 80, 81(14,15), 83(15,17), 87, 89(22), 91, 93(23), 94(23), 102(36), 104(36), 105(38), 106(38), 111(44), 112(48), 115(46), 118(46), 119(44,46,53), 121(44), 122-124, 141(68), 153
Ohsaka, S., 115(52), 124
Ohta, H., 58(40), 71
Ohtani, M., 97(30), 123, 126, 151
Ohuchi, K., 49, 70
Okajima, K., 134(61), 153
Okano, K., 10(6), 23, 25(7, 8), 33(7,8), 35(7,8), 69, 112(46), 115(46), 118(46), 119(46), 124
O'Keeffe, T. W., 164(15), 178
Okunev, A. I., 134(62), 153
Olander, D. R., 66(79,80), 72, 111(43), 124
Ono, S., 126, 151
Orihara, N., 49(32), 70
Orton, B. R., 96(27), 101(34), 123

Oxtoby, D. W., 175(35), 179
Ozawa, S., 20(12), 23, 26(22-27,30), 49(23,25,34), 51(23), 53(34), 56(27), 57(26), 58(34), 62(30, 51), 70, 71, 81(14,15), 83, 85, 87(19), 122, 123

P

Pamies, A., 130(48), 131(48), 152
Pancirov, R. J., 60(44), 71
Papazian, H. A., 132(56), 149, 150, 153, 154
Parry, G., 63(75), 72
Passerone, A., 127(42), 152
Pavlov, V. A., 142(72), 154
Pelipetz, M. G., 3(7), 7, 49(31), 70
Penninger, J. M. L., 60(45), 71
Perlova, N. L., 135(66), 153
Pokrasin, M. A., 135(66), 153
Polukhin, V. A., 175(33), 179
Popel, P. S., 141(70), 142(72), 153, 154
Popel, S. I., 96(24,25), 123
Popova, L. A., 96(24), 123
Poutsma, M. L., 60(43,46), 71
Powell, C. J., 187(17,18,20), 189, 190, 200
Preiswerg, H. P., 176(40), 179
Pulham, R. J., 63(59-63,67-73, 75), 64(70), 71, 72

R

Rabinovich, B. S., 68, 72, 73
Reynolds, C. L., Jr., 142(71), 153
Rice, S. A., 160(8,10), 161, 163(8,10,11,14), 168, 170(11), 171-174, 175(29, 30,36,37), 176(27,38), 178, 179
Rideal, E. K., 145, 154

Robinson, P. J., 58(41), <u>71</u>, 79(9), <u>122</u>
Rodway, D. C., <u>182</u>(9,10), 183(10,11), 185(11), 186(9,11), <u>200</u>
Roshchupkin, V. V., 135(66), <u>153</u>
Rosinberg, M.-L., 127(26), <u>151</u>
Rosner, D. E., 66(78), <u>72</u>
Ruberto, R. G., 83(16), <u>123</u>
Rudin, H., 187(19), 190(<u>19</u>), 193(19), <u>200</u>
Rudnick, L. R., 60(48), <u>71</u>
Ryabov, V. G., 106, <u>124</u>
Ryazanov, M. I., 174(<u>31</u>), <u>178</u>

S

Saito, Y., 9(1,2), 10(3-6), 11(7), 15(1), <u>23</u>, 25(1-4,6-8,11,<u>12</u>,15), 26(6,19), 30(1-3), 31(1,4), 33(1,7,8), 35(1,7,8), 36(1,8), 37(19), 46(1,11), 47(1, 11), <u>69</u>, 75(3,4), 78(3, 4), <u>91</u>, 93(23), 94(4), 95(4), 97(4), 101(4), 102(4,36), 104(4,36), 105(4,38), 106(4,38), 122-<u>124</u>, 141, <u>153</u>
Sako, H., 134(61), <u>153</u>
Sangiorgi, R., 127(<u>42</u>), <u>152</u>
Sasaki, K., 62(51), <u>71</u>, <u>81</u>(15), 83(15), <u>123</u>
Savvin, V. S., <u>134</u>, <u>153</u>
Schlosberg, R. H., 60(<u>44</u>), <u>71</u>
Schultz, L. G., 161(12), <u>178</u>
Schwab, G. M., 3, 4, <u>7</u>, 22, <u>23</u>, 25(9), <u>69</u>
Schwanke, A. E., 136(58), <u>153</u>
Seah, M. P., 181(3), <u>200</u>
Semenchenko, V. K., 101(35), 103(35), <u>123</u>, 126, 127, 141, <u>151</u>
Shah, Y. T., 83(16), <u>123</u>
Shaw, B. A., 96(27), <u>123</u>

Shebzukhov, A. A., 194(35,36), <u>201</u>
Sheppard, N., 52(35), <u>70</u>
Shimamura, N., 62(50), <u>71</u>
Shimoji, M., 125(3), 126, 133, 142(3), <u>150</u>, <u>153</u>, 155(3), <u>177</u>
Sholokhov, V. M., 134(62), <u>153</u>
Siekhaus, W. J., 66(79,80), <u>72</u>, 111(43), <u>124</u>
Siskin, M., 60(47), <u>71</u>
Siskind, B., 161(11), 163(11), 170, <u>178</u>
Slius, D., 173(28), 176(38), <u>178</u>, <u>179</u>
Smith, N. V., 158, <u>177</u>
Soda, H., 127(34), <u>130</u>(34), 148, <u>152</u>
Solbrig, C. W., 12, <u>23</u>
Somorjai, G. A., 141(<u>69</u>), 142(69), <u>153</u>, 173, <u>178</u>, 181(2), 194(29), 196, 199, <u>201</u>
Spann, J. R., 81(13), <u>122</u>
Spicer, W. E., 182(4), <u>200</u>
Starling, K. E., 133(53), <u>153</u>
Steacie, E. W. R., 2, <u>6</u>
Stepanova, N. V., 127, <u>151</u>
Stone, J. P., 81(13), <u>122</u>
Storch, D., 127(19), <u>151</u>
Storch, H. H., 3(7), <u>7</u>, 49(31), <u>70</u>
Strong, S. L., 96(28), 97(28), <u>123</u>
Stroud, D., 127(27,29), <u>152</u>
Suenaga, T., 26(30), 62(<u>30</u>), <u>70</u>, 81(14), <u>122</u>
Sugawara, H., 25(<u>14</u>,<u>16</u>), 46(16), <u>69</u>, 87(21), 88(21), <u>89</u>(22), 91(21), <u>123</u>
Sugimori, A., 62(50), <u>71</u>

T

Takahashi, K., 17(11), <u>23</u>, 25(10), 26(17,18), 31(10), 33(10), 36(17,18),

[Takahashi, K.] 37(19,21),
  43(21), 49(10), 69, 70,
  79(11,12), 80(11,12),
  119(11,53), 121(11),
  122, 124
Taylor, H. S., 2, 6
Tewari, B. N., 134, 153
Thompson, J. C., 193, 200
Thompson, M., 193(27), 201
Timofeyer, A. I., 134(63),
  141(63), 153
Tombrello, T. A., 194(32),
  197(32), 199(32), 201
Tong, S. Y., 193(26), 201
Trigger, S. A., 127(28), 152
Tsukamoto, H., 107(42),
  109(42), 110(42), 124,
  193(34), 196, 201
Tueting, D., 60(48), 71
Tyson, I. F., 193(27), 201

U

Ukhov, V. F., 175(33), 179
Unezhev, B. K., 127(37), 152

V

Van Hook, W. A., 79(10), 122
Vaselow, R., 193(26), 201
Vernon, L. W., 60(42), 71
Versluis, K., 60(45), 71
Vosko, S. H., 164(15), 178

W

Wagner, C. N. J., 97(32,33),
  123
Ward, M. J., 127(43), 143(43),
  152
Waseda, Y., 97(30), 123, 126,
  131, 151, 153
Watabe, M., 127, 151
Weller, R. A., 194(32),
  197(32), 199(32), 201

Weller, S., 3, 7, 49, 70
Wender, I., 65(77), 72
White, D. W. G., 127, 130(45),
  133(45), 148, 152
Whitehurst, D. D., 58(38), 70
Whittingham, A. C., 63(74), 72
Williams, F. L., 142(73), 154
Williams, G. I., 96(27), 123
Williams, G. P., 182(7-10),
  183(10,11), 185(11),
  186(8,9,11), 197, 200
Williams, M. W., 176(39), 179
Wilson, E. G., 160(10),
  161(10), 163(10), 178
Winborne, D. A., 66, 72
Wood, D. M., 127(27), 152
Wotherspoon, J. T. M., 183(12),
  200
Wullschleger, J., 182(6), 200

Y

Yamase, O., 143(78), 154
Yamazaki, H., 26(26), 57(26),
  70
Yan, D., 127(35), 130(35), 152
Yokoyama, T., 9(2), 23,
  25(12), 69, 102(36),
  104(36), 105(38), 106(38),
  123, 124, 141(68), 153
Yoshida, H., 9(2), 23, 25(12),
  69, 102(36), 104(36),
  105(38), 106(38), 123,
  124, 141(68), 153
Young, W. H., 127(20), 151
Yuan, W., 68(84), 73

Z

Zadumkin, S. N., 115(49), 124,
  127(37,39), 152,
  187(15), 200
Zalotai, L., 68(82,83), 72, 73
Zamyatin, V. M., 141(70),
  142(72), 153, 154
Zhukova, L. A., 96(25), 123

Subject Index

A

Accommodation coefficient, 68
Acetone, 31, 33, 46
Acetylene, 64
Acrylonitrile, selectivity for, 35
Activation
  energies, 33, 58
  surface bimolecular, 78
Active center, 2
Activity coefficient, 103, 106, 141
Admolecules
  activated, 76
  long, 148
Adsorbate, 116, 117, 145
  hydrogen atom in, 116
  molecule, 115, 116, 143, 145
  occupied levels of, 116
Adsorption, 62-64, 75, 76, 126, 143, 145
  amount of, 103, 143, 145
  of atoms, 143
  of 2-butanol, 106
  energy of, 106
  enthalpy of, 145
  entropy of, 145, 147
  equation, 145
  equilibrium constant, 14, 15, 17, 77, 105, 144
    for reactant alcohol, 75

[Adsorption]
  heat of, 77, 78, 144
  isotherm, 144, 145, 147
    Langmuir type of, 76
  layer, 173-175
  mobile, 77, 78, 145
  models, 141
  monolayer, 141
  onto sample liquid, 102
  sites, 144
  of solute molecule, 147
  statistical theory of, 77
  system, 145
AES, 107, 196
  analysis, 111
  method, 197
Ag
  4d peak, 186
  surface enrichment of, 198
Ag-Au-Cu, 135
Ag-Cu, 186
  UPS study of, 197
Ag(liq.), 126, 159, 163
Ag(liq.)-O_2, 144
Al, 185, 190
  soft X-ray emission study of, 185
Al-Bi, 134
Al-Cu, 141
Alcohols
  aromatic, 31, 125
  dehydrogenation of, 26, 30, 33, 75, 91, 111, 115

211

[Alcohols]
  octadecyl, 145
  polyhydric, 31
    reactivities of, 31
  primary, 30
  secondary, 30
  tertiary, 31
  unsaturated, 31
Aldehydes
  saturated, 31
  unsaturated, 31
Aldimine, 35
Alkali metals, 37, 65
  in liquid state, 23, 63
  molten, 65
Alkylamines, 23
Alkylbenzenes, 37
Al(liq.), 25, 26, 30, 141, 175
  activity of, 33
  reaction of, 66
Alloy composition, 94, 95
  change in, 195
  as a function of, 115
Allylamine, 35
Alumina, 159
Aluminum, 26
Amines, 33
  aromatic, 36
  dehydrogenation of, 33, 119
  lower, 25
  unsaturated aliphatic, 35
Ammonia, 33
Amplitude, complex, 161
Angle
  critical, 173
  of incidence, 158-160, 176
  of reflection, 188
Angular frequency, 156
Apparatus, gas chromatographic, 17
Ar, 66, 134, 135
  + H_2, 135
  ion bombardment, 196
  ion sputtering, 109
Arrhenius plots, 89
As, 134
Asphaltenes, 51, 83, 87
  concentration of, 85

[Asphaltenes]
  formation, 51, 53, 55
  hydrogenation of, 53, 56
Atomic arrangement, short-range ordering of, 184
Atomic number, 149
Atomic orbital, 115
ATR, 171
Attack
  axial, 87
  equatorial, 87
Attenuated total reflection, 171
Au, 193, 198
  5d, 193
  4f, 191
  $4f_{5/2}$, 198
  $4f_{7/2}$, 198
  pure, 199
Auger
  electrons, 197
  electron spectroscopy, 107, 196
  peak, 197
  spectroscopy, 199
    scanning, 199
Au(liq.), 126, 163
Au-Sn
  electronic structure of, 193
  XPS studies of, 191
Autoclave, 20
  inner wall of, 22

B

Ba, 64
Backscattering, of Auger electrons, 197
Band
  models, 193
  structure, 181
Benzene, 81
Benzylamine, 33, 36
o-Benzyl phenol, 81
Benzyl phenyl ether
  decomposition of, 62, 81
  total conversion of, 82

Subject Index

Bi, 43, 95, 190
  EELS study of, 189
  electronic structure of, 189
Bibenzyl, 81
Binary alloys, 65
  catalyst, 65
  Hg-based, 4
  Tl-based, 4
Bi(liq.), 62, 81, 173
  EEL spectra for, 189, 190
Binding energy, 192
  of Au 4f, 191
  for Hg 5d level, 186
  for Sn 3d, 191
Bi-Pb-Hg, 134
Bloch-Rice model, 170
Bonds, multiple, 65
Borneol-menthone system, 91
BR-model, 170
  skewed, 170
Bulk
  conductivity, 170
  modulus, 149
1,3-Butadiene, 37
  conversion to, 68
  yield of, 43
1,3-Butandiol, 31
2,3-Butandiol, 31
n-Butane, 64
2-Butanol, 49, 78
  adsorbed, 106
  adsorption, 106
  adsorption of, 106
  vapor, 102, 104
    in the presence of, 106
Butene, dehydrogenation of, 37
di-n-Butylamine, 33
γ-Butyrolactone, 31
Buoyancy, 4, 5, 9, 11

C

Cadmium, supported, 2
Cage effect, 62
Camphor, 91
Capillary depression method, 134

Carbon, 196
Carbonaceous materials, 5, 10
Cartesian coordinates, 12, 171
Catalyst
  fouling, 5
  preparatory tube, 9
  purification, 9, 10
  vapor, 17, 19, 20
CD, 134
Cd, 2
  addition of, 141
Cd(liq.), 19, 62, 95
  as catalyst, 81
Cd-Zn system, 96
Cesium bromide, 161
Chamber, gas-tight, 163
Charge transfers, 186
  s-, 186
Chemical potential, 135
Chemisorption, of fenchol, 91
Chloride, 67
  monolayer of, 67
Citronellal, 46
Citronellol, 46
  dehydrogenation of, 46
Cl_2, 66, 68
  molecular beam, 68
Close-packing, 185
CMA, 107, 197
Co(liq.), 126, 163
Coals, 20, 57, 87
  concentration, 85
  concentration of, 85
  conversion, 58
  fragment radicals of, 56
  model compound of, 81
  structure, 58
  of various ranks, 49
Coal liquefaction, 26, 52, 62, 83
  activities, 58
  catalyst, 3
  chemistry, 60
  effect in, 57
  efficient, 53
  over liquid metal, 20
  products, 51
  with Sn (liq.), 49
Co-area
  of admolecule, 77

[Co-area]
  of adsorbate molecule, 145
  magnitude of, 148
  of oxygen, 143
Collision, bimolecular, 75, 76, 87
Compounds, optically active, 88
Compressibility, isothermal, 131
Contaminants, 9
Contamination, 5, 130, 158, 196
  of catalyst surface, 15
  elements, 196
  optical observation of, 22
  due to oxygen, 133
Core
  electrons, 191, 192
  level, 191
Correlation, volcano-shaped, 58, 62
Corrosion, 22
Critical temperature, 148
Cr-Kα, 173
Cs atom, 175
Cs(liq.), 130, 133, 148, 159
  surface of, 175
Cs-Tl, 193
Cu, 186, 198
  3d peak, 186, 198
Cu (liq.), 130, 133, 148, 159
  -S, 144
  surface tension of, 126
Cumene, 37, 46
Cycloalkanol, 31, 33
Cyclohexane, 145
Cyclohexanone, 47
  monosubstituted, 88
Cyclohexene, 68
Cyclohexylamine, 33
p-Cymene, 37, 46

D

Deactivation, 5
Decomposition, 31, 43, 81
  of benzyl phenyl ether, 62, 81

[Decomposition]
  of formic acid, 4
  petroleum, 4
  surface unimolecular, 76, 106
Dehydrogenation, 31, 37, 80
  of adsorbed 2-butanol, 106
  of alcohols, 26, 30, 75, 111, 115
    active catalysts for, 36
  of aliphatic hydrocarbons, 26
  of amines, 33, 119
  of butene, 37
  of citronellol, 46
  of ethylbenzene, 37
  of hydrocarbons, 25, 36, 111
  of isobutanol, 31
  of methanol, 93
  of polynuclear aromatic hydrocarbons, 25, 119, 122
  pyrolytic, 79
  quantum chemical study on, 115
  selective, 35
  selectivity of, 31, 33
  studies, 25
  of tetralin, 36, 79, 80
  of unsaturated aliphatic amines, 35
Delocalization of electron, 116, 117
Density
  distribution, 171, 172, 175
    electronic, 171
    ionic, 172
  electronic, 171
  oscillation, 173, 174
  profile, 173-175
    oscillatory, 176
Density of state, 176
  electronic, 181
Deuterium
  atoms, 79
  tracer techniques, 47
Deuteroalcohol, 79
Dialkylamines, 33
Dielectric constant, 156, 158

Subject Index

Dielectrics, 160-163
Differential grid method, 12
Diffusion
  coefficient, 13
  of gases, 109
  through metallic part, 131
Dihydro-naphthalene, 79
Direct current conductivity, 157
Discharge, electric, 158
Distance
  atomic, 149
  interatomic, 119
Distribution
  atomic, 96
  electronic, 172
    profile, 174
  ionic, 133, 171, 172
Distribution function
  partial, 101
  radial, 97, 101, 119, 131
DR-model, 174-176
Drude-like, 168, 176
  non-, 168, 176
Drude theory, 156, 157, 163

E

EDC, 176, 181
  fine structures of, 184
  peak, 186
  peak shift, 186
  of pure Ag, 186
EEL
  spectra, 190
  of Bi (liq.), 189
EELS, 187, 189, 190
Electromagnetic wave, 155, 156
Electrons
  collective motions of, 188
  delocalization of, 116, 117
  density, 157, 172
  density distribution, 172
    oscillating, 171
  diffraction, 96
  -donating ability, 111, 119

[Electrons]
  donating power, 65
  energy, 181
  energy loss, 190
  energy loss spectroscopy, 187
  gun, 107
  interaction energy, 119
  kinetic energy, 181
  mean free path, 181
  property, 111, 112
  quasi-free, 172
  spectroscopic technique, 180
  spectroscopy, 180, 181
  transfer, 63
    successive, 64
    from tetralin, 120
Electron delocalization
  energy, 115, 116
  from metal, 116
  through H(2) atom, 116
  from the highest occupied level, 118
Electron theory, of metals, 4
Electronic properties, 157
  of liquid metals, 4, 156
  of metal, 149
Electronic states, 191
Electronic structures, 168, 197
  of Al, 185
  of Au-Sn alloy, 193
  changes in, 190
  of inner shells, 181
  of liquid metals, 157, 187
Ellipsometer, 157
Ellipsometry, 157
Enantiomer differentiating reaction, 89
Energy
  distribution curve, 176
  levels, 68
  of adsorbate, 115, 116
Enthalpy
  of adsorption, 145
  of oxide formation, 58
Entropy, of adsorption, 145
Environment, atomic, 192

Equation of states, 133
Ethane, 64
Ethanol, 47
  -In(liq.), 78, 79
Ether linkages, 58
Ethylbenzene, 37, 81
  dehydrogenation of, 31
Ethylene, 64
  glycol, 31

F

F_2, 66
Factor, geometrical, 17, 20
FD-MS, 52
Fe-As, 134
Fe(liq.), 126, 143
  optical properties of, 163
Fe(liq.)-O_2, 144
Fe(liq.)-S, 144
Fe(liq.)-S-C, 144
Fe(liq.)-Se, 144
Fenchol
  -camphor system, 91
  chemisorption of, 91
  d-, 88, 89
Fe-Ni binary alloy, 135
FeO, 143
Fermi
  energy, 116
  level, 115, 117, 176, 181, 186
Flow-type, 10
  apparatus, 12
Formaldehyde, 30
Formic acid, decomposition, 3, 4
Fowler's equation, 127, 133
Freedom
  internal, 58
  of reacting molecule, 58
Free electron theory, 157, 190
Free energy, of adsorption, 106
Free radicals, 26, 60
  organic, 58
Frequency
  factor, 33

[Frequency]
  vibrational, 77
Furan, 4
Furfural, 4

G

Ga, 185
  bulk concentration of, 199
  electronic structure of, 186
  solid, 185
  at the melting point, 185
Ga-In-Sn, 135
Ga(liq.), 25, 26, 33, 35, 62, 81, 130, 159
  EELS studies of, 190
Gallium, 30
Gas, two-dimensional, 145
Ga-Te, 193
Gaussian
  distribution function, 171
  function, 174
Gibbs
  adsorption equation, 140-145
  dividing plane, 168
Gibbs-Duhem relation, 135
GPC, 52
GR-model, 171, 172

H

Hard sphere diameter, 133
Heat
  of adsorption, 77, 78, 144
  of fusion, 62
Helium, 19, 20
Hg
  5d, 186
  $5d_{3/2}$, 186, 190
  $5d_{5/2}$, 186, 190
  sites, 186
Hg(liq.), 172, 175
  EELS studies of, 190
  optical conductivity of, 170

Subject Index

[Hg(liq.)]
  optical constants of, 161
  optical properties of, 168, 170, 172, 176
  photoemission study of, 176
  surface of, 172
  transition zone of, 171, 173
  work function of, 175
  X-ray reflection intensity for, 173
Hindrance, steric, 87-89, 91, 93
Huckel method, extended, 119
Hydrocarbons
  aliphatic, 26
    dehydrogenation of, 26
  aromatic, 26
    unsaturated, 49
    dehydrogenation of, 25, 36, 111
  polynuclear aromatic, 65
    dehydrogenation of, 25, 119, 122
    hydrogenation of, 65
Hydrogen, 91, 102
  acceptors, 49
  atoms, 64, 115, 116, 119
    adsorbed, 64
  donor, 36
  purification device, 9
  purified, 9, 158
  -shuttling solvent, 58
  transfer reactions, 43, 47, 49, 91
    between alcohols and ketones, 25, 87
    between optically active compounds, 88
    over liquid metal catalyst, 87
    utility of, 47
Hydrogenation, 65
  of asphaltenes, 53, 56
  of polynuclear aromatic hydrocarbons, 65
Hydroxy citronellal, 47
Hydroxy citronellol, 47
Hyperbolic tangent function, 173

I

Impurities, 5, 10, 131
In, 43, 67, 68, 95, 190
  atoms, 101, 186
    incorporation of, 186
    charge transfer from, 186
  content, 110
  surface concentration of, 104, 105
  surface enrichment of, 199
  surface segregation of, 110, 111
In-Al, EELS studies on, 190
In-Bi, 190
  liquid alloy, 101
    catalytic activity of, 106
Incidence, oblique, 161
In-Cl, 67, 68
Indium, 26, 67
  -methanol system, 119
  -tellurium system, 119
Inert gas, 10, 163
Infrared region, 160
In-Ga, eutectic alloy of, 199
In-Hg, liquid alloy, 170, 180
In(liq.), 25, 26, 62, 78, 79, 81, 93, 120
  catalyst, 87
    surface of, 47, 89
  fenchol over, 89
  high selectivity of, 30
In-Pb
  liquid alloy, 101
    catalytic activity of, 106
  surface of, 196
In-Sn solid alloy, at the melting point, 111
In-Te, 193
Interband excitations, 187, 190
Ion bombardment, 196
Ionization potential, 111
  first, 111, 119
  of Te(liq.), 122
Ion scattering spectroscopy, 199
Ion-sputtering technique, 109
IR, 52

Isobutanol, dehydrogenation of, 31
Isomer
  d-, 89, 93
  ℓ-, 89, 93
  steric, 91
Isomerization, 43, 62
  pathway, 62
  rate of, 62
Isotope effects, 79
  kinetic, 78, 79
ISS, 199

K

K, 25, 26, 37, 65, 111
K_2CO_3, 65
Ketones, 25, 30, 31, 43, 46, 87, 91
  effect of, 47
  molecule, 47
  unsaturated, 49
Khafji crude, 62
Kinetics, 75, 78, 79
  irreversible first order, 33, 81
  Langmuir type, 17
  of reactions, 12
K(liq.), 19, 64
  experimental values for, 127
  reactivity of, 64
KOH, 65
Kramers-Kronig relation, 161

L

Langmuir
  adsorption equation, 145
  adsorption isotherm, 144, 145
  equation, 145
Langmuir-type
  adsorption equation, 76
  kinetics, 17
  rate equation, 14
Large-drop method, 134

Layers
  atomic, 172, 175, 197
  topmost, 175, 197, 199
LCAO-MO, 115
Lead, 67
  reaction of, 4
LEED, 172, 173
Levels
  occupied, 116, 118
  unoccupied, 116, 118
Li
  activity of, 65
  dispersion of, 65
Li(liq.), 63, 64, 127
Lindemann's theory, 149
Liquefaction
  Sn(liq.)-catalyzed, 53
  of coal, 49
Liquid alloys
  binary, 37, 95, 101, 134, 135
  catalytic activities of, 94, 112
  PES studies of, 186
  under hydrogen atmosphere, 144
  Pt-based, 141
  surface tension of, 134
  ternary, 134
Liquid zinc, contaminated by oxygen, 3
Lithium fluoride, 161
Long-range order, of atomic arrangement, 181
Low energy electron diffraction, 172

M

Maximum bubble pressure method, 101, 127, 134
MBP, 134, 135
Mean free path, of Auger electrons, 197
Measurements
  ellipsometric, 158, 168
  reflectometric, 168

Melting points, 2, 9, 26, 67, 68, 108, 110
  of catalysts, 65
Melting temperature, 26
Menthol, 94
  d-, 94
  dehydrogenation of, 93
  dℓ-, 93, 94
  ℓ-, 94
Menthone, 89, 93
  adsorbed, 91
  d-, 89
  ℓ-, 88
Mercury, 135, 145, 168, 170, 190
  optical property of, 171
  surface, 148
  surface tension of, 145
Metals, noble, 163
Metal-semimetal, transition, 193
Methanol, 31, 33
  decomposition, 2, 95
  dehydrogenation, 30, 120
  indium-, 119
  -In(liq.), 78
  tellurium-, 119
Methylcyclohexanone, 87
1-Methylnaphthalene
  H-shuttling effect of, 58
  -tetralin mixed solvent, 57
MgF_2, 163
Mg-Kα, 191
Mg(liq.), 127
Microfeeder, 10
Mixing cup concentration, 13, 15
Models
  catalyst, 119
    energy curve for, 120
  structural, 53
Molecular beam, 66, 68, 111
Molybdenum, 159
Monolayer, 67
  admolecules, 145
  adsorption model, 141
  capacity, 144
Monte Carlo simulation, 173, 175
Multicomponent system, 134, 135
Multilayer, 145, 146

N

Na, 26, 37, 65, 111
Na-Cs, 64, 65
  liquid alloy, 175
Na(liq.), 19, 64
  bulk of, 64
  experimental values for, 127
  transition zone of, 175
Na-Rb, 65
Naphthalene, hydrogenation, 65
Neomenthol, 93
Ni(liq.), 126, 163
Nitriles, 33
Nitrobenzene, 2
NMR, 52
Noble gas ion, 199
Nonmetal, 119
Normal incidence, 160, 161, 193
Normal reflection, 160, 161, 163

O

Octadecane, 145
Oils, 56
  fractions, 86
  -1, 51, 83, 87
    concentration of, 85
    yield of, 55
  -2, 51, 83, 87
    concentration of, 85
    yield of, 55
Optical axis, 160
Optical cells
  designs of, 161
  glass made, 158
Optical conductivity, 157, 168, 170, 171
Optical constants, 156, 160, 163
  of Hg(liq.), 161

[Optical constants]
  magnitude of, 160
Optical properties, 156, 163, 171
  ellipsometric, 176
  of Hg(liq.), 168, 170, 172
  of liquid metals, 155, 156
  of mercury surface, 171
  reflectometric, 176
  of simple metal, 156
Oscillating drop method, 130
Overlapping, d-d, 186
Oxidation, 97, 163
  method, 196
  susceptibility to, 30
Oxides
  of catalyst, 9
  catalytically unfavorable, 26
  impurities such as, 10
  layer, 2
    activity of, 3
Oxygen, 109, 131, 196
  adsorbed by Fe(liq.), 143
  amounts of, 115
  on the surface, 111
  atoms, 25, 58
  Co-area of, 143
  contaminants, 131
  incorporation, 130

P

Pair
  correlation function, 126
  interaction potential, 126
  primary, 62
Parameters
  kinetic, 33
  structural, 53, 86
Partition function, for vibrational motion, 77
Pb, 43, 67, 95, 135, 196
  surface of, 68
  vaporization of, 196
$PbCl_2$, 68
Pb-Sn, 134
PE, spectra, 185

Perfumes, 46
Perturbation potential, 116
PES, 181
  studies of liquid alloy, 186
Petroleum
  decomposition, 4
  residues, 26
Phase
  adsorbed, 147
  surface, 197
Phenol, 62, 81
Photoelectrons, 181
Photoemission, 176
  spectroscopy, 181
Photon energy, 176
Plasma frequency, 157
Plasmon frequency, 150
Plasmon losses, 188
p-Polarized light, 157
Polymerization, 56
  of fragment radicals, 51
Potential energy surface, 79
Preasphaltenes, 51, 83
  concentration of, 85
Prism, 163
Probe depth, 199
Products, random, 81, 83
Profile
  electronic, 175
  ionic, 175
  monotonic, 175
2-Propanol, 78
  -In(liq.), 79
Propylbenzene, 81
Propyne, 63
Pseudoatom theory, 173
Pt, 141
Pt-Al, 134
Pt-Co- 134
Pt-Pd, 134
Pt-Si, 134
Pt-Sn, 134
Pulse
  reactor, 16
  size, 17
Pump, electromagnetic, 23
Pyrolysis, 20, 80

Subject Index

R

Radicals
  $C_2H_5\cdot$, 64
  fragment, 51
  of coal, 56
  polymeric organic, 62
  reactions, 58
  reaction scheme for, 81
Rate
  equation, 84, 85, 87
    analysis of, 89
    experimental, 76
    initial, 16
Rate constant, 13-15, 17, 58, 75, 78, 82, 85, 86
  for j-isomer, 89
  observed, 58
  of surface unimolecular reaction, 106
  values of, 80
Rayleigh's equation, 130
Rb(liq.), 127
Reaction
  model, 87
  pathways, 93
  rate, 75, 91
  scheme, 62, 75, 76, 83, 87, 93
    complexity of, 91
    for radical reaction, 81
Reactor
  aerodynamic levitation, 66
  bubbling, 10, 11
  melt bed, 4
  rectangular duct, 12
    modified, 17
    use of, 14
Rearrangement, 62
Reflection
  amplitude, 157
    complex, 159
  angle, 173
  coefficient, 161
  optical, 193
  phase change on, 157
Reflectometry, 157, 166, 167
Refraction, double, 163

Refractive index, 156, 160, 173
Regular solution, monolayer, 142
Relaxation time, 157
Restored azimuth, 157, 158
Rigid sphere model, 133

S

S, 130, 143
Sb, 135
Schiff's base, 33
SD, 134, 135
Se, 37, 135, 143
Selenium, 37
  ionization potential of, 122
Semimetal, 19, 25, 36
Sessil drop method, 127, 134
Sharp surface model, 168
Shin-Yubari coal, 49
  liquefaction of, 53
Side reactions, 5, 10
  forming Schiff's base, 33
  initiation of, 35
Sintering, 5
Sites, multiplet, 80
Small-angle X-ray reflection, 176
Small-angle X-ray reflectometer, 173
Sn, 43, 95, 101, 105, 191
  3d, 191
  4d, 191
  3d electrons, 191
  pure, 199
  5s, 192
  spectral intensities for, 198
Sn-Al, surface composition of, 196
Sn(liq.), 3, 49, 51, 56, 58, 62, 81, 86, 159
  activity of, 86
  catalyst, 49, 55-57, 62, 86
  coexistence of, 87

[Sn(liq.)]
  diffraction pattern of, 173
  pure, 197
  surface of, 56, 68, 69
Sn(II) oxide, 69
Sodium, 112
  chloride, 161
  work function of, 113
Solid solution, with low
    melting point, 122
Solvent, hydrogen-donating,
    55
Soxhlet extraction, 51
Specimen holder, horizontal,
    193
Sputtering yield, 196
ss-Model, 176
Standard enthalpy, of oxide
    formation, 58
State, pseudocrystalline, 185
Statistical theory, of adsorption, 77
Stearic acid, 147
Steel, 159
Step function, 172, 175
  profile, 175
Stereochemistry, 87
  of catalysis, 47
Stirring
  of liquid metal, 11
  magnetic, 23
  during reaction, 20
Structure, open, 185
Sulfur, 130
Surface
  analyses, 101, 181, 197
  of In-Sn liquid alloy, 107
  of liquid alloys, 193, 196, 198
  area, 14, 22, 106, 140, 145
    molecular, 147
    partial molar, 141
  chemistry, 26
  chloride, 67
  clean, 109, 111
  complex, 105
  composition, 101, 104, 106,
    110, 115, 135, 140-142,
    181, 186, 193, 197

[Surface]
  [composition]
    of Al-Cu liquid alloy, 141
    analysis of, 181
    determination of, 126
    determined by AES, 197
    of In-Bi liquid alloy, 106
    of In-Pb liquid alloy, 106
    of In-Sn liquid alloy,
      103, 107, 111
    of liquid alloy, 111, 196
    of Sn-Al alloy, 196
    of Sn-Au liquid alloy, 199
  contamination, 130, 163,
    180, 196
    elimination of, 196
    by oxygen, 109
  coverage, 144
  enrichment, 198, 199
    of In, 199
  entropy, 133, 134
    of liquid metal, 130
  excess, 103, 143
  free, 107, 163
  impurities, 143
  layer, 147
  pressure, 145, 147
  reactions, 63
  segregation, 110, 135, 143,
    144, 170
    of Ag, 186
    of In, 110, iii
    of Pt-based binary liquid
      alloy, 141
    zone, depth of, 170
  structure, 131, 174
    anomalous, 133, 157
    of Hg(liq.), 175
    special, 172
  transition zone, 168
  plasmon, 190
    dispersion, 171
    loss peak, 189
    loss spectra, 189
Surface tension, 101-103,
    106, 115, 126, 127,
    130, 131, 133, 135,
    140, 141, 144, 148,
    149, 197

[Surface tension]
  advanced theories of, 127
  of binary liquid alloys, 134
  of Bi-Pb-Hg liquid alloy, 134
  cell, 131
  correlation of, 148
  of Cu(liq.), 130
  depression of, 130
  equation, 126
  information from, 131
  of In-Sn liquid alloy, 102, 141
  of liquid alloys, 106, 134
  of liquid metal, 143, 144, 150
  measurements, 105, 106
  at the melting point, 148
  of mercury, 145
  of multicomponent liquid alloys, 135
  studies, 63, 101
  temperature dependence of, 133, 142, 149
  theory of, 126

T

Te, 135, 143
  on Fe(liq.), 143
  particles, 81
  vapor, 79, 80
    catalysis, 81
Te(liq.), 19, 25, 36, 79, 119, 120
  catalysis by, 80, 119
  catalysis of, 119
  catalyst, 119
  catalytic activity of, 37, 119
  electron-donating ability of, 119
  intrinsic activity of, 80
  ionization potential of, 122
  surface of, 80
Tellurium, 199, 122

[Tellurium]
  atom, 121
  ionization potential of, 121
  -methanol system, 119
  -tetralin system, 119
Terpene alcohols, 25, 46
  unsaturated, 31
    dehydrogenation of, 31
Te-Se
  catalytic activity of, 37
  liquid mixture, 122
Tetrahydrofuran, 31
Tetralin, 36, 53, 55, 57, 80, 87
  dehydrogenation, 36, 79, 80, 120
  indium-, 119
  1-methylnaphthalene-, 57
  solvent effect of, 55
  tellurium-, 119
Thallium, 30
Thallium bromide-iodide, 161
Theory of adsorption, statistical, 77
Thermal expansion, 149
Thermodynamics, statistical, 145
Ti, 144
Ti-Fe, 144
Tin
  liquid, 3
  shots, 49
Tl(liq.), 25, 26, 93
  high selectivities of, 30
Tl-Te, 193
Toluene, 65, 81
Transition state, 62, 88
  formation of, 87
  models, 79
Transition zone, 168, 170-175
  advanced studies on, 171
  of Hg(liq.), 171, 173
  model, 170, 171, 173
  of Na(liq.), 175
  nonoscillatory, 174
  oscillatory, 174
  with special structure, 176
  width of, 131, 168, 170, 172, 173, 177
Tungsten, 159

U

UHV, 66, 180
  apparatus, 186, 193
  cell, 176
  chamber, 181
  conditions, 190
Ultraviolet photoelectron
  spectroscopy, 197
Unit, optical, 163
UPS, 197

V

Vacuum
  chamber, 159, 163
  ultrahigh, 66, 180
  ultraviolet, 158, 163
Valence-band, 192
Valence electrons, 191, 192
Vaporization, heat of, 148
Vapor pressure, 19, 26, 30, 67, 197
Volmer-type, adsorption equation, 145
Volume plasmon, 190
  loss, 188
  peak, 189, 190
  spectra, 189

W

Wave function, 116
Windows, 161, 163
  slit-like, 22
Work function, 112, 113, 117, 118, 175, 186
  hypothetical, 112
  of liquid metal, 113
  of polyvalent liquid metal, 187
  of sodium, 113

X

XP
  process, 199
  spectra, 198
    for core electrons, 191, 192
  spectral data, 197
  spectral intensities, 199
XPS, 190, 191, 198
  studies, 191, 193, 197
    of Au-Sn
X-ray
  emission, soft, 185
  photoelectron spectroscopy, 190
  reflection, 172-174
    coefficient, 174
    intensity, 173, 174, 176
  reflectometer, small angle, 173
  scattering, 96, 101, 126, 131
    intensities of, 97
  total reflection of, 173

Z

Zinc, 111
  liquid, 30
  metallic, 2, 30
Zn, 94
  incorporation of, 95
Zn(liq.), 19, 25, 33, 35
  catalyst, 36
  as catalyst, 81
  catalyzed by, 26
  experimental values for, 127
  selectivity of, 31
Zn-Bi, 134
Zn-Pb, 134

RAYMOND H. FOGLER LIBRARY